AF885314

A Critical Study of the Law of Increasing Entropy

- *its fallacious nature*
- *its correction*
- *the emergence of negentropic universes*
- *the logical and dialectical perspectives*

José Cruz Íñiguez Romero

INFINITY PUBLISHING

All rights reserved. No part of this book shall be reproduced or transmitted in any form or by any means, electronic, mechanical, magnetic, photographic including photocopying, recording or by any information storage and retrieval system, without prior written permission of the publisher. No patent liability is assumed with respect to the use of the information contained herein. Although every precaution has been taken in the preparation of this book, the publisher and author assume no responsibility for errors or omissions. Neither is any liability assumed for damages resulting from the use of the information contained herein.

Copyright TXu 1-776-447 © 2012 by José Cruz Íñiguez Romero
Front Cover: Original design by the author

ISBN 978-0-7414-7768-2

Printed in the United States of America

Published August 2012

INFINITY PUBLISHING
1094 New DeHaven Street, Suite 100
West Conshohocken, PA 19428-2713
Toll-free (877) BUY BOOK
Local Phone (610) 941-9999
Fax (610) 941-9959
Info@buybooksontheweb.com
www.buybooksontheweb.com

To my wife:
Beatríz Ivette Ibarra Lemas,

To my daughters:
Barbara and Argentina,

To my grandsons:
Nathan and Noah

To the memory of Dr. Eduardo Césarman
(1931-2004)
A true renaissance man

ACKNOWLEDGEMENTS

The author wishes to express his appreciation and gratitude to the following Publishing Houses for granting him the permission to quote copyrighted material essential to the discussions herein presented.

The McGraw-Hill Companies permitted the author to quote from the following works
1. K. S. Pitzer and L. Brewer (Revision of Lewis and Randall) *Thermodynamics*, Second Edition, © 1961 by McGraw-Hill, Library of Congress Catalog Card Number 60-10604
2. Gordon M. Barrow, *Physical Chemistry*, Third Edition, © 1973 by McGraw Hill, ISBN 0-07-003823-6
Lawrence & Wishart, Ltd., kindly allowed the author to quote from Volume 25 of *Karl Marx and Frederic Engels Collected works*, ISBN 0-7178-0525-5 (v. 25)

Dover Publications Inc. granted the author permission to quote from the following works:
1. Sadi Carnot: *Reflections on the Motive Power of Fire*, Copyright © 1960 by Dover Publications, Inc.
2. Max Planck: *Treatise on Thermodynamics*, Third Edition, ISBN 0486-66371-X, Library of Congress Catalog Card Number 45-9816
3. Ernst Schmidt: *Thermodynamics*, Copyright © 1966 by Dover Publications, Inc., Library of Congress Catalog Card Number 66-22160

Elsevier Publishing Co., authorized the author to quote from Ivan P. Bazarov's *Thermodynamics*, Pergamon Press, 1964, Library of Congress Control Number 63015495

CONTENTS

CHAPTER 5: CURRENT SECOND LAW THERMODYNAMICS

CHAPTER 6: THE FLAWS IN CURRENT SECOND LAW THERMODYNAMICS

CHAPTER 7: THE CORRECT VALUES FOR THE TRANSFORMATIONS AND THE EMERGENCE OF THE NEGENTROPIC FORMULATION

CHAPTER 8: IRREVERSIBLE CYCLICAL PROCESSES

SYNOPSIS

Four are the essential pieces of new knowledge advanced in this work.

1. An irrefutable thermodynamic argument is presented to negate the supposed equivalence between that principle of experimental truth embodied in the statement "Heat cannot, of itself, pass from a colder to a hotter body", and the notion stating that "The entropy of the universe tends to a maximum"

2. A body of thermodynamic evidence is presented to show that Clausius' version of the second law is a formulation essentially distinct from current second law thermodynamics. The said evidence sustains the conclusion that the latter is nothing more than the product of a correction performed on Clausius' analysis with the deliberate purpose of salvaging the law of increasing entropy.

3. Detailed logic and dialectical analysis of Clausius' formulation as well as current second law thermodynamics allows unveiling the contradictions at the core of these two formulations.

4. The properly corrected version of Clausius' work developed by this author *–the negentropic formulation-* shows that the entropy change for the universe of a given thermodynamic process is determined by the interplay of entropic, irreversible processes such as the direct flow of heat from high to low temperature or the degradation of work into heat, and negentropic, reversible processes such as the necessarily reversible transformation of heat into work. The predominance of the former kind of processes leads to entropic universes, the predominance of the latter to negentropic universes, and the perfect balance between them to non-entropic universes.

CHAPTER 1: Clausius, Engels, and Aristotle

1.1 ANTECEDENTS

The first attacks on Clausius' law of increasing entropy from the perspective of dialectics were those of Engels. They read as follows:

Clausius—if correct—proves that the universe has been created, ergo that matter is creatable, ergo that it is destructible, ergo that also force, or motion, is creatable and destructible, ergo that the whole theory of the 'conservation of force' is nonsense, ergo that all his conclusions are nonsense [Engels, 1987, p. 562]

Clausius second law, etc. however it may be formulated, shows energy as lost, qualitatively if not quantitatively. Entropy cannot be destroyed by natural means but it can certainly be created. The world clock has to be wound up, then it goes on running until it arrives at a state of equilibrium from which only a miracle can set it going again. The energy expended in winding has disappeared, at least qualitatively, and can only be restored by an impulse from outside. Hence, an impulse from outside was necessary in the beginning also, hence, the quantity of motion, or energy, existing in the universe was not always the same, hence energy must have been created, i.e. it must be creatable, and therefore destructible. Ad Absurdum! [Engels, 1987, p. 563]

Engel's is certainly not alone in his conviction. Teilhard de Chardin [1967], Roger Caillois [1973], and M. L. McGlashan [1966] to name a few, have also expressed from their own particular perspectives, their reservations about this law, the law of increasing entropy, which others, like Eddington [1915], hail as the supreme law of Nature.

The inquiry to be conducted in this work will approach Clausius' work on the second law from a thermodynamic perspective, using as analytical tools both: logic, as well as the three fundamental laws of dialectics.

1.2 INTRODUCTION

Having as guide the notions establishing that:

- the unity of the world consists in its materiality
- motion and matter are indivisible, indestructible and uncreatable
- the infinitely multiform phenomena of nature are only various forms of the motion and development of matter

Engels managed to turn Hegel's dialectics upside down, replacing its idealistic footing for one rooted on experience. In rejecting Hegel's essential notion that ideas are superior to matter, and in recognizing matter and motion as the ultimate components of reality, Engels established dialectics as *…the science of the general laws of motion and development of nature, human society and thought…* [Engels, 1987, p. 131]; acknowledging, in the process, the fundamental difference between the metaphysical and dialectical approaches to the study of natural phenomena. To metaphysics… *things and their mental reflexes, ideas, are isolated, are to be considered one after the other and apart from each other, are objects of investigation fixed, rigid, given once and for all…* [Engels, 1987, p.22]; *…Dialectics, on the other hand, comprehends things and their representations, ideas, in their essential connection, concatenation, motion, origin, and ending…*[Engels, 1987, p.23].

Dialectics: this expanded outlook of nature through which *…all rigidity was dissolved, all fixity dissipated, all particularity that had been regarded as eternal became transient*, and in which *…the whole of nature was shown as moving in eternal flux and cyclical course…* [Engels, 1987, p. 327] finds support in the three laws of dialectics, as follows [Engels, 1987, p. 356]:

1. The law of transformation of quantity into quality and vice versa

2. The law of the interpenetration of the opposites
3. The law of negation of the negation

1.3 THE LAW OF THE TRANSFORMATION OF QUANTITY INTO QUALITY AND VICEVERSA

The law of the transformation of quantity into quality and vice versa, can be expressed by saying that *…in nature, in a manner exactly fixed for each individual case, qualitative changes can only occur by the quantitative addition or quantitative subtraction of matter or motion (so-called) energy…* [Engels, 1987, p. 357]. This law can be clearly exemplified by the qualitative changes inherent to the phase transitions [ice / liquid water / water vapor] brought upon in one direction or the other by the addition or subtraction of heat. These qualitative changes go beyond the simple differences in physical appearance of the different forms of water, into the realization that the emergence of each new qualitatively different state brings forward or demand new or different laws for their description.

1.4 THE LAW OF THE INTERPENETRATION OF THE OPPOSITES

This law is the reflection of the fact that both, the being and the becoming of matter encompass opposite elements in a permanent state of tension: protons and neutrons in the atom; positive and negative in electricity, magnetism, and mathematics; attraction and repulsion forces defining, among other things, the physical states of matter, etc.

The particular condition matter happens to adopt under a given set of circumstances is the expression of the particular balance adopted under those circumstances by the opposites at play. The upsetting of this balance by quantitative changes leads to another state of balance between the opposites. Change in matter is nothing more than the reaction of the contradiction expressed by the opposites to the new stimuli i.e. the resolution of the new tensions created in it by the said quantitative changes in matter or motion. Thus understood, the cause of change is inherent to matter itself i.e. resides within it, and not in that 'alien' impulse from outside referred to by Engels in the second of the quotes offered at the beginning of this paper; impulse from outside that can be found playing a central role in alternative looks at nature and its development.

1.5 THE LAW OF THE NEGATION OF THE NEGATION

In Engels words:

Negation of the negation...(is) a very simple process taking place everywhere and every day" [Engels, 1987, p. 125]...Let us take a grain of barley. Billions of such grains of barley are milled, boiled and brewed and then consumed. But if such a grain meets with conditions that are normal for it, if it falls in suitable soil, then under the influence of heat and moisture it undergoes a specific change, it germinates; the grain as such ceases to exist, it is negated, and in its place appears the plant which has arisen from it, the negation of the grain. But what is the normal life-process of this plant? It grows, flowers, is fertilized and finally once more produces grains of barley, and as soon as these have ripened the stalk dies, is in its turn negated. As a result of this negation of the negation we have once again the original grain of barley, but not as a single unit, but ten-, twenty- or thirty fold. Species of barley change extremely slowly, and so the barley of today is almost the same as it was a century ago. But if we take a plastic ornamental plant, for example a dahlia or an orchid, and treat the seed and the plant which grows from it according to the gardener's art, we get as a result of this negation of the negation, not only more seeds, but also qualitatively improved seeds, which produce more beautiful flowers, and each repetition of this process, each fresh negation of the negation, enhances this process of perfection [Engels, 1987, p.126]...It is the same in mathematics. Let us take any algebraic quantity whatever: for example, a. If this is negated we get – a (minus a). If we negate that negation, by multiplying – a by – a, we get $+a^2$*, i.e. the original positive quantity, but at a higher degree, raised to its second power... [Engels, 1987, p. 127].*

The fact that...*negation of the negation expresses the connection of the old and the new, and the repetition at a higher stage of development of some properties of the lower stage of a process...* [MIA, Encyclopedia of Marxism], leads to the representation of development, of progress, as an expanding three-dimensional helix: the lower point reappears, but at a higher level.

1.6 FORMAL LOGIC OR ARISTOTLE'S PRINCIPLE OF CONTRADICTION

According to Aristotle:

"He who knows best about each genus must be able to state the most certain principles of his subject, so that whose subject is existing things qua existing must be able to state the most certain principle of all things. This is the philosopher, and the most certain principle of all is that regarding which it is impossible to be mistaken; for such a principle must be both the best known (for all men may be mistaken about things which they do not know), and non-hypothetical. For a principle which every one must have who understands anything that is, is not a hypothesis; and that which every one must know who knows anything, he must already have when he comes to a special study. Evidently then such a principle is the most certain of all; which principle this is, let us proceed to say. It is, that the same attribute cannot at the same time belong and not belong to the same subject and in the same respect; we must presuppose, to guard against dialectical objections, any further qualifications which might be added. This then, is the most certain of all principles, since it answers to the definition given above. For it is impossible for any one to believe the same thing to be and not to be, as some think Heraclitus says. For what a man says, he does not necessarily believe; and if it is impossible that contrary attributes should belong at the same time to the same subject (the usual qualifications must be presupposed in this premises too), and if an opinion which contradicts another is contrary to it, obviously it is impossible for the same man at the same time to believe the same thing to be and not to be; for if a man were mistaken on this point he would have contrary opinions at the same time. It is for this reason that all who are carrying out a demonstration reduce it to this ultimate belief; for this is naturally the starting point even for all other axioms.

There are some who, as we said, both themselves assert that it is possible for the same thing to be and not to be, and say that people can judge this to be the case. And among others many writers about nature use this language. But we have now posited that it is impossible for anything at the same time to be and not to be, and by this means have shown that this is the most indisputable of all principles. Some indeed demand that even this shall be demonstrated, but this they do through want of education, for not to know of what things one should demand demonstration, and of what one should not, argues want of education. For it is impossible that there should be demonstration of absolutely everything (there would be an infinite regress, so that there would still be no demonstration); but if there are things of which one should not demand demonstration, these persons could not say what principle they maintain to be more self evident than the present one"[Aristotle, 350 BCE]

Having defined our analytical instruments, we will proceed to establish in the next chapter, the thermodynamic framework within which our analysis is to be conducted.

CHAPTER 2: BACKGROUND

2.1 CARNOT AND THE MOTIVE POWER OF FIRE

Most of the foundation of the field that was to become known as second law thermodynamics was laid by Carnot in his 1824 memoir titled *Reflections on the Motive Power of Fire and on Machines Fitted to Develop that Power*. In it he tackled -within the confines of the caloric theory of heat- the problem of determining the factors affecting the work generating capability of heat engines. The problem was worded by him as follows:

The question has often been raised whether the motive power of heat is unbounded, whether the possible improvements in steam engines have an assignable limit –a limit which the nature of things will not allow to be passed by any means whatever; or whether, on the contrary, these improvements may be carried on indefinitely. We have long sought, and are seeking today, to ascertain whether there are agents preferable to the vapor of water for developing the motive power of heat; whether atmospheric air, for example, would not present in this respect greater advantages. We propose now to submit these questions to a deliberate examination. [Carnot, pp. 5-6]

A comparison between work production in heat engines with that taking place by harnessing the potential energy of waterfalls led him to the realization that paralleling the latter, the motive power of heat also depended on the quantity of heat flowing and on what he called "the height of its fall, that is to say, the difference of temperature of the bodies between which the exchange of caloric is made" [Carnot, p. 15].

In regard to the influence of this difference of temperatures he further elaborated as follows

In the waterfall the motive power is exactly proportional to the difference of level between the higher and lower reservoirs. In the fall of caloric the motive power undoubtedly increases with the difference of temperature

between the warm and cold bodies; but we do not know whether it is proportional to that difference. We do not know, for example, whether the fall of caloric from 100 to 50 degrees furnishes more or less motive power than the fall of this same caloric from 50 to zero. It is a question that we propose to examine thereafter. [Carnot, p.15]

The problem addressed by Carnot in this last quote is that of the efficiency of heat engines. Even if as discussed below, he was capable of understanding the way this magnitude behaved along the temperature scale, the development of a functional relationship –an equation- between efficiency and temperature eluded him. The development of such a relation had to wait for Clausius to resume the work of Carnot. It serves to say that in doing so Clausius not only developed an equation for work production efficiency in terms of the temperatures of the hot and cold reservoirs, an equation which actually constitutes a discriminating criterion for the operations of cyclical processes; Embedded in this equation he recognized a state function, the entropy -intimately related to the efficiency question- which his analysis embodied with the capability of discriminating or separating cyclical operations in two groups: reversible, or not. It should also be stated in regard to the discussions that follow that what Carnot is referring to when speaking about "the difference of temperatures" is a temperature difference between two specific temperatures, i.e. a temperature gradient at a specific location in the temperature scale.

Having deferred the efficiency question to a later stage in his analysis, Carnot's immediate goal was reformulated as follows:

It is natural to ask this curious and important question: Is the motive power of heat invariable in quantity, or does it vary with the agent employed to realize it as the intermediary substance, selected as the subject of action of the heat?
It is clear that this question can be asked only in regard to a given quantity of caloric, the difference of the temperatures also being given...We will try to answer these questions, availing ourselves of ideas already established. [Carnot, p. 9]

Getting at the answer of the previous question demanded from Carnot the invention of what Mendoza calls "the closed cycle of operations" [Carnot, p. xii] now known as Carnot's engine or Carnot's cycle, as well as recognition of the fact that in such a cycle the maximum of work production can only be achieved if:

...there is produced on the body no other change of temperature than that due to the change of volume, or what is the same thing in other words, if there is no contact between bodies of sensibly different temperatures [Carnot, p. 20]

A situation to be avoided on reason of the fact that

...contact between bodies of different temperatures (brings) loss of motive power [Carnot, p. 14]

The previous notions, it should be recognized constitute the operative conditions respectively associated with reversibility and irreversibility.

With these antecedents in place Carnot further realized that just like work could be obtained from the flow of caloric from the hot to the cold reservoir, work needed to be expended to effect the transfer of caloric in the inverse direction. Furthermore, he realized that if in both of these operations there had been no loss of either motive power or caloric (if both operations were reversible) then

...The quantity of motive power produced in the first place will be equal to that which would have been expended in the second, and the quantity of caloric passed in the first case from the body A to the body B would be equal to the quantity which passes back again from the body B to the body A [Carnot, p. 11]

This same conclusion was later voiced by Carnot through a statement that constitutes the immediate antecedent to Planck's reversibility criterion as quoted below in Section 2.2

...The result of these first operations has been the production of a certain quantity of motive power and the removal of caloric from the body A to the body B. The result of the inverse operation is the consumption of the motive power produced and the return of caloric from the body B to the body A; so that these two series of operations annul each other, after a fashion, one neutralizing the other. [Carnot, p. 19]

The previous statements, referring to the coupling of two reversible cycles, one engine and one refrigerator, subsume the notion that under the same conditions of operation no engine can be more efficient than a reversible engine. Its veracity was proved by Carnot via a *reductio ad absurdum* argument using as basis of proof the impossibility of creation of motive power out of nothing:

...If there existed any means of using heat preferable to those we have employed, that is, if it were possible by any method whatever to make the caloric produce a quantity of motive power greater than we have made it produce by our first series of operations, it would suffice to divert a portion of this power in order by the method just indicated to make the caloric of the body B return to the body A from the refrigerator to the furnace, to restore initial conditions, and thus be ready to commence again an operation precisely similar to the former, and so on: this would be not only perpetual motion, but an unlimited creation of motive power without the consumption of either caloric or of any other agent whatever. Such a creation is entirely contrary to ideas now accepted to the laws of mechanics and of sound physics. It is inadmissible [Carnot, pp. 11-12]

The just proved fact that for a given amount of heat flowing and a given temperature difference no particularity introduced in the operation of a reversible engine –change of working substance included- can have any improving effect in its work production capability was eventually re-expressed by Carnot in the form now known as Carnot's Theorem:

The motive power of heat is independent of the agents employed to realize it; its quantity is fixed solely by the temperatures of the bodies between which is effected, finally, the transfer of caloric [Carnot, p. 20]

At this point on his analysis, most of the foundation of the science of heat engines –a parcel of knowledge that through Clausius' efforts was to become known as second law thermodynamics- had been laid. All the pieces of the puzzle were there, except one: The functional relation between the efficiency of the operation and the temperatures of the reservoirs. Even if Carnot was capable of recognizing the essence of this functional relationship, its actual mathematical form eluded him. Carnot understood that:

> *...the fall of caloric produces more motive power at inferior than at superior temperatures* [Carnot, p. 36]

Beyond this he recognized that even if

> *...We have shown that the quantity of motive power developed by the transfer of caloric from one body to another depends essentially upon the temperature of the two bodies...we have not shown the relation between these temperatures and the quantities of motive power produced* [Carnot, p. 35]

Concluding with:

> *...We are not prepared to determine precisely, with no more experimental data than we now posses, the law according to which the motive power of heat varies at different points on the thermometric scale* [Carnot, p. 37]

The accomplishment of this task: the mathematical definition of the reversible efficiency, among others, was reserved for Clausius. He introduced two critical modifications into Carnot's work. He abandoned the caloric theory of heat underwriting Carnot's analysis and chose a different basis of proof for his results. The first of these modifications brought Carnot's analysis into compliance with Joule's results regarding the inter-conversion of heat and work. The work of Joule and others was essential to the final acceptance of the principle of the conservation of energy, or as Clausius would name it: the first fundamental principle of the mechanical theory of heat. From this perspective, Carnot's assumption concerning the equality of the amounts of heat that in a cycle of operation the variable body receives from the hot reservoir and gives out to the cold reservoir, had to be abandoned. In regard to Carnot's position Clausius recognized that:

...This mode of dealing with the question does not accord to our present views, inasmuch as we rather assume that in the production of work a corresponding quantity of heat is consumed, and that in consequence the quantity of heat given out to the surrounding space during the cyclical process is less than that received by it. Now if for the production of work heat is consumed then...there is no ground for saying that the work is created out of nothing. Accordingly not only must the principle enunciated by Carnot receive some modification, but a different basis of proof from that used by him must be discovered [Clausius, 1879, p. 77]

This new basis of proof he found in that "statement of experimental truth" as Schmidt calls it [Schmidt, 1966, p. 92] which in embodying the essential message of the second law can be considered its foremost expression:

Heat cannot, of itself, pass from a colder to a hotter body. [Clausius, 1879, p. 78]

Armed with these notions he was able, in 1850, first to reword and then to prove the correctness of Carnot's theorem in its new form. The proof carried on by Clausius, in essence identical to that *reductio ad absurdum* argument previously used to that same end by Carnot, showed that denying the validity of Carnot's theorem implied the unassisted transference of heat from a cold to a hot body, result that stands in direct opposition with experience. Clausius rewording of Carnot's theorem took the following form:

...the relation between the quantity of heat carried over, and that converted into work, is independent of the nature of the matter which forms the medium of the change [Clausius, 1879, p. 79]...(and only depends) *on the temperatures of the two bodies... which act as heat reservoirs* [Clausius, 1879, p. 81]

It should be noted that Carnot's original reference to "the motive power of heat" has been replaced in Clausius' re-expression by "the relation between the quantity of heat carried over, and that converted into work". In doing so

Clausius changed the emphasis from the amount of work produced to work production efficiency, as it is efficiency what is defined by the said relation.

In order to understand how was it that Clausius solved the efficiency problem and from it transited to the entropy function and, eventually, to the law of increasing entropy, we need to make sure we understand the operation of Carnot's reversible cycle, not in its original caloric-centered version, but in that which accords to the dictates of the law of conservation of energy. This is the matter of the section that follows

2.2 THAT WHICH IS KNOWN BEYOND ANY DOUBT ABOUT REVERSIBLE CYCLICAL PROCESSES

The discussions that follow will make constant reference to a reversible cyclical process, also known as a heat engine or as Carnot's cycle, such as that represented in Figure 1. The central role played by this contraption in the work herein presented is a consequence of the fact that the work of Clausius leading to the formulation of the second law of thermodynamics was developed using this reversible cyclical process as a model for the inter-conversions of the energy forms known as heat and work. Therefore, any argument directed to criticize the work of Clausius or the tenets of what we now understand as second law thermodynamics' should begin with a clear description of the operation of this thermodynamic cycle. This is the objective of the present section.

The constitutive elements of Carnot's cycle, such as that schematically represented in Figure 1, are the following

1. A hot reservoir of temperature T_h
2. A cold reservoir of temperature T_c, with $T_h \rangle T_c$

3. A working substance, also known as the variable body, which for the purpose of this discussion will be taken to be an ideal gas, and
4. A mechanical reservoir (not shown) to act as depository for the work output of the cycle. The initial condition of this device indicates that no work is deposited in it.

For the purpose of the discussions that follow, the initial conditions of the heat and mechanical reservoirs are understood as those these bodies have before the operation of the engine takes place.

One cycle in the operation of this engine may be described via the following concatenation of processes. This description, which makes reference to Figure 1, is not the only one possible. An alternative description, differing in the starting point, is presented in Section 4.1

1. Process *AB*: An isothermal and reversible expansion at the temperature (T_h) of the hot reservoir. Here an amount of heat Q_h absorbed by the ideal gas from the hot reservoir, is transformed into an equivalent amount of work W_h, i.e. $Q_h = W_h$
2. Process *BC*: An adiabatic and reversible expansion through which the ideal gas manages to produce an amount of work W' out of its own internal energy, and in doing so its temperature drops from T_h to T_c.
3. Process *CD*: An isothermal and reversible compression at the temperature (T_c) of the cold reservoir. The work W_c expended in carrying on this process is released as an equivalent amount of heat Q_c to the cold reservoir, i.e. $Q_c = W_c$ It should also be noted here that $W_c \langle W_h$.
4. Process *DA*: An adiabatic and reversible compression which brings the cycle to its completion. This process

demands the expenditure of an amount of work W' identical to that produced by process *BC*. Due to the adiabatic nature of the process, the work expended ends up increasing the internal energy of the ideal gas, thus raising its temperature from T_c to T_h. The reader should note here that this process manages to revert or cancel all the changes initially produced by process *BC*.

The information just presented will be now used to inquire about the effects produced by one cycle in the operation of this reversible cycle. In this regard it will have to be agreed from the start that whatever form these effects might take they won't include the variable body, this on reason of the fact that at the end of one cycle such a body returns to its precise original condition. Being this so, these effects are to be searched for in the other three bodies in such a cycle participating, namely, the heat and mechanical reservoirs. The already mentioned fact about the changes produced by process *BC* being precisely cancelled or reverted by those of process *DA* allows us to realize that the cycle's effects should be solely traceable to those produced by processes *AB* and *CD*. As previously described, through isothermal and reversible expansion *AB* the ideal gas manages to absorb an amount of heat Q_h from the hot reservoir, and transform it into an equivalent amount of work W_h which ends up in the mechanical reservoir. Along *CD*, on the other hand, a portion W_c of this work is retrieved from the mechanical reservoir and used to carry on the indicated isothermal and reversible compression *CD*. The spent work ends up in the cold reservoir as the equivalent amount of heat Q_c. This heat-to-work/work-to-heat concatenation just described represents, in simpler terms, the transference of the amount of heat Q_c from the hot to the cold reservoir. The fact that the accomplishment of this effect demands only the degradation of the amount of work W_c means that the remaining portion of W_h, in the amount of $W_h - W_c$, remains as such in the mechanical reservoir. This remnant, the net work output of the cycle, will be from now on designated as W, i.e. $W = W_h - W_c$. If Q is now defined as $Q = Q_h - Q_c$ then, and on reason of the facts that $Q_h = W_h$ and $Q_c = W_c$ we will also have that $W = Q = Q_h - Q_c = W_h - W_c$.

The previous findings can be summarized by saying that out of the energy received from the hot reservoir in the form of Q_h, the concatenation of processes *AB*, *BC*, *CD*, and *DA*, manage to effect two changes

- The transfer of an amount of heat Q_c from the hot to the cold reservoir, and
- The production of an amount of work equal to the difference between Q_h and Q_c.

Both, the detailed and the summarized renditions of this cycle have been depicted in Figure 1.

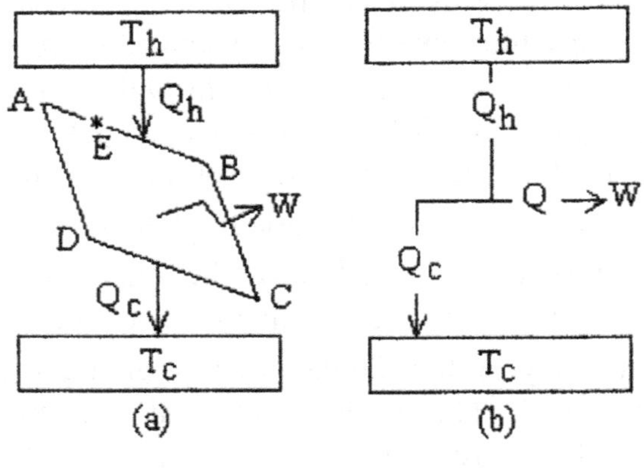

Figure 1. Two representations of a heat engine: In (a) the cyclical evolution defined by the concatenation of processes AB-BC-CD-DA allows the variable body (an ideal gas) to absorb an amount of heat Q_h from the hot reservoir; discard an amount of heat Q_c to the cold reservoir, and transform the difference $Q = Q_h - Q_c$ into an equivalent

amount of work W. Figure (b) is a schematic representation of the two changes or transformations brought upon by the described cycle in the operation of engine 1(a). Point E is to be used in the identification and calculation of the values or entropy changes for the two transformations shown in 1(b).

It seems necessary to point out here that the reversible qualifier ascribed to the processes of Figure 1 stems from the facts that the work-producing circuit *ABCDA* receives and processes all of the heat released by the hot reservoir, as well as releases all of the heat absorbed by the cold reservoir. In doing so it manages to produce the maximum possible amount of work at the conditions of operation. If any amount of heat coming out of the hot reservoir bypasses in any way possible the work-producing circuit and flows directly to the cold reservoir, then the work producing potential of this heat is wasted, and if so the work output of the cycle will be less than the maximum in precisely the amount of work that this amount of heat would have rendered if entering the cycle. In such a situation we would have to recognize three as the effects of the cycle: The transformation of a certain amount of heat into work; the transformation of another quantity of heat from the hot to the cold reservoir via concatenation *ABCDA*; and the direct transfer –bypassing *ABCDA*- of still another amount of heat from the hot to the cold reservoir. This matter is further discussed in the following paragraphs.

The evolution of the variable body –the ideal gas- in a direction opposite to that shown in Figure 1(a) defines a reversible refrigerator. To understand the functioning of this inverse cycle, and to get at the same time a deeper understanding of the reason behind the reversible qualifier ascribed to cycles such as that represented in Figure 1, let us assume that the operation of the engine has just been completed so that the changes previously described for the heat and mechanical reservoirs are in effect. In other words, the inverse cycle will be carried on using the same heat and mechanical reservoirs and in the condition they were left at the conclusion of the engine's operation above described.

Let us now subject the ideal gas to the following concatenation of processes:

1'. Process *AD*: An adiabatic and reversible expansion through which the ideal gas manages to produce an amount of work W' out of its internal energy, and in doing so its temperature drops from T_h to T_c. At the end of this process the mechanical reservoir is in possession of an amount of work $W + W'$. The amount of work W being that previously deposited in it by the occurrence of the engine's *ABCDA* cycle.

2'. Process *DC*: An isothermal and reversible expansion at the temperature of the cold reservoir. Here an amount of heat Q_c absorbed by the ideal gas from the cold reservoir is transformed into an equivalent amount of work W_c. At the end of this process the cold reservoir ends up releasing an amount of heat identical to that received during the engine's operation, recuperating this way its precise original condition. The mechanical reservoir, on its part, safe keeps now an amount of work $W + W' + W_c$.

3'. Process *CB*: An adiabatic and reversible compression carried on via the expenditure of an amount of work W' identical to that produced by process *AD*. Due to the adiabatic nature of the process, here the work expended ends up increasing the internal energy of the ideal gas, raising its temperature from T_c to T_h. At the end of this process the amount of work deposited in the mechanical reservoir has changed to $W + W' + W_c - W'$, an amount of work the reader should have no difficulty in identifying as equal to W_h.

4'. Process *BA*: This isothermal and reversible compression at the temperature of the hot reservoir demands an amount of work W_h. Let us here recognize the fact that at this point the inverse cycle has only managed to produce an amount of work W_c which by itself is incapable of carrying on compression *BA* to completion. If this inverse cycle is to be completed then the missing work, in the amount of $W = W_h - W_c$ should come from somewhere else. In the case at hand, where the concatenation engine-refrigerator is taking place, this work is that deposited in the mechanical reservoir by the initial operation of the engine, which combined with W_c, can carry on the inverse cycle to completion. The spent work W_h is finally absorbed by the hot reservoir which this way recuperates both, the heat by it released during the engine's operation, as well as its initial condition. At the end of compression *BA* we also find the mechanical reservoir in its original condition, as the total amount of work in it deposited at the end of *CB*, has been withdrawn and expended along *BA*. This concatenation makes clear two things. First that while the engine managed to produce work out of the heat flowing from the hot to the cold reservoir; the refrigerator, on its part, consumes work in transferring heat from the cold to the hot reservoir; Second, that the reversible refrigerator cancelled i.e. reversed each and every change produced by the engine in a way such that at the end of the coupling all the bodies involved, i.e. the heat and mechanical reservoirs (the universe of the process) have returned to their precise initial conditions, without changes remaining elsewhere in the universe. This means plain and simple that the condition of the universe before the engine's operation is *identical* to the condition it is found at the conclusion of the coupling engine/refrigerator.

The reversible condition –the possibility of recuperating the initial state of the universe without changes remaining elsewhere- can be achieved via the fulfillment of two stringent (actually impossible) requirements: The operation has to be propelled by infinitesimal gradients and every source of dissipation –say friction- has to be eliminated. The fact that none of these conditions can actually be accomplished makes reversible processes the efficient, ideal limit of natural processes. Any real engine working between the same two heat reservoirs will actually output, from the same amount of heat Q_h, a smaller amount of work than its reversible counterpart, and precisely because of this, the recuperation of the initial condition of the universe will be impossible without changes remaining elsewhere. To illustrate in a very simple manner this point let us define the efficiency η of a heat engine in terms of the amount of work w it manages to produce out of a given amount of heat Q_h: $\eta = w / Q_h$. Of all the engines operating within given heat reservoirs, the reversible engine is the one that manages to transform the largest chunk of Q_h into work; in other words, any real engine is less efficient than a reversible engine. The less efficient of them all is that engine incapable of producing any work out of Q_h. In this situation $w = 0$, and consequently $\eta = 0$. In this particular operation, which serves well to illustrate the concept of reversibility, all of the heat released by the hot reservoir ends up in the cold reservoir after bypassing the work-producing circuit. Recuperation of the initial condition of the universe in this case demands the transfer of an amount of heat Q_h from the cold to the hot reservoir, without changes being left elsewhere. This amount of heat can indeed be transferred from the cold to the hot reservoir via a refrigerator. The problem is that the operation of the refrigerator demands work, and given that the initial process produced none, we have none at our disposal. We can indeed import from somewhere else the amount of work needed to transfer Q_h from the cold to the hot reservoir. Doing this, however, does not recuperate the initial condition of the universe. The reason being that at the end of it all we will realize that the final universe is different than the original in at least the change experienced by the body called to supply the missing work. The only possibility left to recuperate the initial condition of the universe without changes remaining elsewhere demands of Q_h the capability of flowing by itself, unassisted, from the cold to the hot reservoir. The fact that this possibility is denied by experience means plain and simple that in the case at hand the initial condition of the universe is not recoverable, and because of this the inefficient operation being considered is irreversible. It is only in that case where the engine manages to produce the maximum possible amount of work that the recuperation of the initial condition of the universe is possible. The argument given can be extended to any non-reversible process. These notions have been summarized by Planck as follows

A process which can in no way be completely reversed is termed irreversible, all other processes reversible. That a process may be irreversible, it is not sufficient that it cannot be directly reversed... The full requirement is that it be impossible, even with the assistance of all agents in nature, to restore everywhere the exact initial state when the process has once taken place... " [Planck, p. 84]

It was Clausius whom originally recognized that one cycle in the operation of a reversible cyclical process could be described in terms of the two changes by it left in the universe. These changes or effects, such as those represented in Figure 1(b), were by him called 'transformations'. In his own words...

In a Simple Cyclical Process two variations in respect to heat take place, viz. that a certain quantity of heat is converted into work (or generated out of work), and another quantity of heat passes from a hotter into a colder body (or vice versa)...For the former of these two variations we have already employed the word 'transformation', inasmuch as we said, when work was expended and heat thereby produced, or conversely when heat was expended and work thereby produced, that the one had been 'transformed' into the other. We may use the word 'transformation' to express the second variation also (which consists in the passage of heat from one body into another, which may be colder or hotter than the first), inasmuch as we may say that heat of one temperature 'transforms' itself into heat of another temperature...On this principle we may describe the result of a simple cyclical process in the following terms: Two transformations are produced, a transformation from heat into work (or vice versa) and a transformation from heat of a higher temperature to heat of a lower (or vice versa). The relation between these two transformations is therefore which is to be expressed by the second Main Principle [Clausius, 1879, pp. 91-92]

It was with the aid of the 'transformation' concept that Clausius carried on the essential part of his analysis on the second law.

Let us note that the last sentence of Clausius previous quote identifies the relation of these two transformations as the subject matter of the second law. This statement finds explanation in the fact that just as Carnot before him; Clausius understood that the essence of this field was centered on the transformation of heat into work, a transformation quantifiable, measurable, through the efficiency concept. The way the efficiency is related to these two transformations, one of which refers to the portion of the incoming heat transformed into work, with the other portion passing through to the cold reservoir, can be understood by realizing that any change in the magnitude –in the relative amount- of these portions embodies a change in the efficiency of the operation.

For those used to calculate or describe the total or universe entropy change in terms of the combined entropy changes of the bodies there involved, this 'transformation' concept might sound strange. The interesting part about the use of 'transformations' in the entropy-centered analysis of a given process stems from the fact that this approach perceives thermodynamic processes in terms of their effects, i.e. in terms of the discernible changes processes leave in the universe. As already described, the universe of a reversible cyclical process can be approached in terms of the two changes it leaves in the universe: the transfer or transformation of a certain amount of heat from the hot to the cold reservoir, and the transformation of another quantity of heat into work. On this perspective, the universe, say, of an isothermal chemical reaction could be approached in terms of the transformation of reactants into products; the transformation of a certain amount of heat from the reacting system to the thermal bath, or vice versa, and the transformation of a certain amount of heat into work. Once the discernible effects produced by a given process have been recognized, then the bodies involved in each of these transformations are identified. The combined entropy changes of all the bodies involved in a given transformation determines its entropy change. This fact makes a transformation a universe in itself as well as a sub-universe of the larger process of which it forms part. For certain systems this approach is the most efficient possible. Instead of speaking of the combined entropy changes of this, this, this, and that body, we simply speak of the entropy change of this transformation; an example of this, as will be soon be seen, is the reversible transformation of heat from the hot to the cold reservoir which involves two isothermals, two adiabats, as well as the mechanical reservoir.

As previously said, the entropy change for a given transformation is made up of the contributions of all the bodies in it participating. For example, the entropy change for the irreversible transformation of heat from a hot to a cold body is represented as: $\Delta S[Q(T_h) \to Q(T_c)]_{irr} = -\frac{Q}{T_h} + \frac{Q}{T_c}$.

In this expression we can recognize that the first term of the right hand side represents the entropy change for the hot reservoir; with the second term representing the entropy change for the cold reservoir. The combination of these two entropy changes defines this way the entropy change of the said transformation.

CHAPTER 3: CLAUSIUS FORMULATION OF THE SECOND LAW

3.1 CLAUSIUS QUANTIFIES THE EFFECTS OF THE TRANSFORMATIONS

Clausius' first step towards unveiling the equation relating efficiency and the temperatures of the reservoirs consisted in carefully stating the problem at hand. Where Carnot was looking for the law "according to which the motive power of heat varies at different points on the thermometric scale" [Carnot, p. 37], Clausius, having taken the quotient Q_h / Q_c as a measure of the efficiency of the operation, was looking for an equation of the form $Q_h / Q_c = \phi(T_h, T_c)$, in which $\phi(T_h, T_c)$ represented "some function of the two temperatures which is independent of the nature of the variable body" [Clausius, p. 81]. From here he proceeded to select a suitable working substance with which to carry on the analysis leading to the identification of this function:

> *...The circumstance that the function...is independent of the nature of the body offers a ready means of determining this function, since as soon as we have found its form for any single body it is known for all bodies whatsoever...Of all classes of bodies the perfect gases are best adapted for such a determination, since their laws are the most accurately known. We will therefore consider the case of a perfect gas subjected to a cyclical process, similar to that here reproduced [Clausius, 1879, p. 81]*

Available to him were thus the equations describing the behavior of this ideal gas as it experiences the series of processes associated to the reversible cyclical process. It was through these equations that he was capable of identifying the function ϕ.

The already mentioned fact in reference to Figure 1 that the effects of process *BC* are precisely canceled or reverted by those of process DA circumscribes any effect of the cycle to processes *AB* and *CD*, through which heat and work are quantitatively transformed into one another. Thus, while in *AB* the amount of heat Q_h received by the ideal gas from the hot reservoir is transformed into an equivalent amount of work; in *CD* the amount of work used to carry on this compression ends up being transformed into an equivalent amount of heat Q_c and as such, rejected by the ideal gas to the cold reservoir. It is thus the difference (Q) between these amounts of heat, the one quantifying the work output of the cycle (W), i.e. $Q_h - Q_c = Q = W$

Being work production the primary goal of a heat engine, it should be easy to understand that the true measure of its efficiency will have to compare the work output of the cycle with its heat input, or Q/Q_h. Alternative measures of efficiency can be defined according to which heat stream is the emphasis on. Clausius favoring the ratio Q_h/Q_c might have been motivated by the fact that as shown below, it is this precise ratio what the equations used reduce to. These equations, written for one mole of gas and with a slightly different notation to that used by Clausius [1879, pp. 82-83], connect the extent of isothermal and reversible processes *AB* and *CD* with the amount of heat (or work) exchanged with the heat reservoirs (or mechanical reservoir).

$$Q_h = RT_h \ln \frac{V_B}{V_A} \tag{3.1}$$

$$Q_c = RT_c \ln \frac{V_C}{V_D} \tag{3.2}$$

Also available to Clausius were the equations describing adiabatic and reversible processes *BC* and *DA*. Also with a minor notation modification to that used by Clausius [1879, pp. 82-83] the following relation can be written

$$\frac{V_B}{V_A} = \frac{V_C}{V_D} \qquad (3.3)$$

With these expressions at his disposal Clausius proceeded to perform a series of algebraic manipulations similar to the ones described below.

Substitution of equation (3.3) in (3.1), followed by its division by equation (3.2), leads to the following result for the efficiency of a reversible cyclical process:

$$\frac{Q_h}{Q_c} = \frac{T_h}{T_c} \qquad (3.4)$$

Let us add in this regard that this equation subsumes all its other possible efficiency re-expressions. They can be obtained by the simple procedure of replacing either Q_h or Q_c according to $Q_h - Q_c = Q$, and solving for the expression of interest. For example, substitution of Q_c for $Q_h - Q$ followed by some algebra produces the expression commonly identified as the efficiency of these cycles: $Q / Q_h = (T_h - T_c) / T_h$.

The long journey initiated by Carnot finds, through the effort of Clausius, its fruitful conclusion with equation (3.4). The theoretical framework of the science of heat engines was now complete. In order to understand the transition from the science of cyclical processes to the second principle of the mechanical theory of heat, or to the second law of thermodynamics as it was later to be called, we have to focus our attention on two notions which subsumed by the efficiency equation, were called to play an instrumental role in Clausius further work in these matters, a work that was to eventually led him to the law of increasing entropy, and through it to be business of predicting the fate of the universe. Their identification starts by recognizing that (3.4) can also be written as:

$$\frac{Q_h}{T_h} = \frac{Q_c}{T_c} \tag{3.5}$$

And also as

$$-\frac{Q_h}{T_h} + \frac{Q_c}{T_c} = 0 \tag{3.6}$$

The fact that

$$Q = Q_h - Q_c \tag{3.7}$$

Allows re-expressing equation (3.6) in the following form

$$[(-\frac{Q_c}{T_h} + \frac{Q_c}{T_c}) + (-\frac{Q}{T_h})]_{rev} = 0 \tag{3.8}$$

The first of these notions –pertaining the values of the transformations- can be traced to equation (3.8) in which it can be noticed that while its leftmost parenthesis refers exclusively to the amount of heat Q_c exchanged between the reservoirs, the other one refers exclusively to the amount of heat Q of temperature T_h that ends up being transformed into work. Judging from the way Clausius reasoned his way through to what he called the values of the transformations - with value meaning the quantifier of the effect of a transformation, a term that was to be eventually replaced by 'entropy change'- as presented in the quote below, it is only reasonable to assume that the pairing of the previous result – equation (3.8)- with the effects of the two transformations produced by a reversible cyclical process, suggested Clausius that those parenthetical expressions were the actual values of these two transformations.

Clausius started with the decision..."*to consider the transformations of work into heat, and therefore the passage of heat from a higher to a lower temperature as being positive quantities*" [Clausius, 1879, p. 97]. From here he went to recognize that "...*the value of a change from work into heat must be proportional to the quantity of heat generated, and that beyond that it can only depend on its temperature...*" [Clausius, 1879, p. 98]. In regard to the transformation of heat from one temperature to another his reasoning went as follows "...*Similarly, the value of the passage of heat* Q_c *from the temperature* T_h *to the temperature* T_c *must be proportional to the quantity of heat which passes, and beyond this can only depend on the two temperatures...*" [Clausius, 1879, p. 98]. On this matter Clausius further clarified his position by saying "... *the passage of the quantity* Q_c *from temperature* T_h *to* T_c *has the same equivalence value as a double transformation of the first kind, viz. the transformation of the quantity* Q_c *from heat of temperature* T_h *into work, and again out of work into heat of temperature* T_c *...*" [Clausius, 1879, p.101].

The translation of the previous verbal statements into mathematical expressions led him to values for the transformations identical to those hinted by equation (3.8), as shown by the following statement [Clausius, 1879, pp. 100-101]

> *...the generation out of work of the quantity of heat* Q *of temperature* T *has the equivalence value* Q/T *; and the transference of the quantity of heat* Q_c *from temperature* T_h *to temperature* T_c *has the equivalence value* $Q_c\left(\frac{1}{T_c}-\frac{1}{T_h}\right)$

The quotes previously transcribed contain, again, minor changes in notation from that of Clausius.

Clausius reduction of the heat transfer process taking place in a reversible cyclical process to a heat-to-work/work-to-heat concatenation leads, at the light of last value quoted in the previous statement, $Q_c\left(\frac{1}{T_c}-\frac{1}{T_h}\right)$, to the recognition of $-Q_c / T_h$ as the value for the transformation into work of the amount of heat Q_c of temperature T_h, and Q_c / T_c as that associated to the transformation of this amount of work into an equivalent amount of heat Q_c at temperature T_c.

If equation (3.8) was a statement about the values of the transformations, equation (3.6) was, on the other hand, a statement of the fact that the summation of the values of these transformations in a reversible cyclical process amounted to zero. These are the two notions recognized by Clausius in the efficiency equation. These two pieces of knowledge were to play an essential role in the construction of what here will be referred to as Clausius' formulation of the second law.

3.2 THE ENTROPY

With the values of the transformations at his disposal, Clausius proceeded to consider in further detail the question of their equal to zero combination. He extended the thermodynamic analysis of the previous section to reversible cyclical processes similar to that represented in Figure 1, but of increased complexity. For these, in correspondence with equation (3.6), he found that the summation of the values of the transformations there taking place took the form $\Sigma Q / T = 0$ [Clausius, 1879, p. 87]. Later on he tackled the analysis of reversible cyclical processes "...not consisting solely of isentropic and isothermal lines, but altogether general in form" [Clausius, 1879, p. 87] in which "...the taking of heat is simultaneous with changes of temperature..." [Clausius, 1879, p. 88]. For these processes he found that the summation of values was in turn given by the expression $\int\frac{dQ}{T}=0$ [Clausius, 1879, p. 89], about which he had the following to say

> *This equation, which was first published by the author in 1854 (Pogg. Ann. vol. 93, p. 500), forms a very convenient expression for the second main Principle of the Mechanical Theory of Heat, as far as it relates to reversible processes. This principle may be expressed in words as follows: If in a reversible Cyclical Process every element of heat taken in (positive or negative) be divided by the absolute temperature at which it is taken in, and the differential so formed be integrated for the whole course of the process, the integral obtained is equal to zero...* [Clausius, 1879, p. 90]

Recognizing as well that

> *...If the integral* $\int dQ/T$, *corresponding to any given succession of variations of a body, be always equal to zero provided the body returns finally to its original condition, whatever the intervening conditions may be, then it follows that the expression under the integral sign, viz.* dQ/T, *must be the perfect differential of a quantity, which depends only on the present condition of a body, and is altogether independent of the way in which it has been brought into that condition. If we denote this quantity by S, we may put* $dQ/T = dS$, *or* $dQ = TdS$, *an equation which forms another expression, very convenient in the case of certain investigations, for the second main principle of the Mechanical Theory of Heat* [Clausius, 1879, p. 90]

This quotient of reversible heat divided by the absolute temperature at which it is exchanged received from Clausius the name *entropy,* and the symbol *S*. The name comes from the Greek word *τροπη*, meaning *Transformation.*

A comparison of the definition just given for the entropy with Clausius previously quoted 'values' leads to the realization that these two terms are in fact the same thing.

Clausius previous statements mark the conclusion of the transition from the science of heat engines to the second law of thermodynamics. In terms of the values for the transformations the above quoted statement of the second law for reversible cyclical processes took the following form "...in a reversible Cyclical Process the total value of all the transformations must be equal to nothing" [Clausius, 1879, p. 102]

3.3 THE ENTROPY CHANGES FOR CLAUSIUS TRANSFORMATIONS

With base on the discussions/results offered/presented in Sections 3.1 and 3.2 we will proceed to make a summary of the values or entropy changes assigned by Clausius to the two transformations produced by a simple reversible cyclical process such as that represented in Figure 1. In order to do this use will be made of the self-evident notation represented by the brackets written below to designate the reversible transformations of heat into work, work into heat, and heat from one temperature to another.

For the reversible transformation of an amount of heat Q $(Q = Q_h - Q_c)$ of temperature T_h, into an equivalent amount of work W:

1. $\Delta S_{Clausius}\ [Q(T_h) \rightarrow W]_{rev} = -Q/T_h$ (3.9)

For the reversible transformation of an amount of work W into an equivalent amount of heat Q of temperature T_h:

2. $\Delta S_{Clausius}[W \rightarrow Q(T_h)]_{rev} = Q/T_h$ (3.10)

For the reversible transformation of an amount of heat Q_c from the hot reservoir of temperature T_h, to the cold reservoir of temperature T_c:

3. $\Delta S_{Clausius}\,[Q_c(T_h) \rightarrow Q_c(T_c)]_{rev} = -\frac{Q_c}{T_h} + \frac{Q_c}{T_c} = Q_c(\frac{1}{T_c} - \frac{1}{T_h})$ (3.11)

And for the reversible transformation of an amount of heat Q_c from the cold reservoir of temperature T_c, to the hot reservoir of temperature T_h:

$$4.\ \Delta S_{Clausius}[Q_c(T_c) \rightarrow Q_c(T_h)]_{rev} = -\frac{Q_c}{T_c} + \frac{Q_c}{T_h} = Q_c(\frac{1}{T_h} - \frac{1}{T_c}) \qquad (3.12)$$

It serves to note that if this amount of heat were to be transformed between two bodies of essentially the same temperature, then equation (3.11) and (3.12) will reduce to zero, i.e.

$$\Delta S[Q_c(T) \rightarrow Q_c(T)]_{rev} = Q_c(\frac{1}{T} - \frac{1}{T}) = 0 \qquad (3.13)$$

Equations (3.9) through (3.12) are simple re-expressions of Clausius's results quoted in Section 3.1. They have been written for both: a heat engine, and a refrigerator. The already discussed fact that a reversible refrigerator brings forward the inverse of every change produced by the reversible engine means that the effects and values of any of these contraptions are the precise opposite of the other. This fact is reflected in the pairs of equations (3.9)/(3.10) and (3.11)/(3.12).

Since the effect of one cycle in the operation of a reversible cyclical process reduces to those embodied by their associated transformations, it follows that the total entropy change for the cycle being considered, or equivalently the entropy change for its universe, will be given by the combination of the entropy changes of the two transformations there taking place. In doing this we get, in compliance with equation (3.8), the following expression

$$\Delta S_{Clausius}[Cycle]_{rev} = \Delta S[Q(T_h) \rightarrow W]_{rev} + \Delta S[Q_c(T_h) \rightarrow Q_c(T_c)]_{rev} =$$

$$-\frac{Q}{T_h} + (-\frac{Q_c}{T_h} + \frac{Q_c}{T_c}) = 0 \qquad (3.14)$$

As previously noticed, a simple algebraic rearrangement of the previous values leads to:

$$\Delta S_{Clausius}[Cycle]_{rev} = -\frac{Q_h}{T_h} + \frac{Q_c}{T_c} = 0 \qquad (3.15)$$

The transition from equation (3.14) to equation (3.15) sheds light on the fact that in Clausius' construction, combining the entropy changes for the transformations is equivalent to the combination of the entropy changes for the heat reservoirs. The reason is simple. The entropy change for the hot reservoir $(-Q_h / T_h)$ subsumes both: The $-Q / T_h$ entropy change for the transformation into work of the amount of heat Q of temperature T_h, as well as the $-Q_c / T_h$ component of the entropy change of the transformation of the amount of heat Q_c from the hot to the cold reservoir. The other component of this entropy change corresponds to the entropy change of the cold reservoir, Q_c / T_c. The result obtained by combining the entropy changes for the transformations associated to a reversible cyclical process given in (3.14) is thus reproduced by the combination of the entropy changes of the reservoirs shown by (3.15).

If we remove the reversible qualifier from equation (3.15) we will end up with the general equation furnished by Clausius' formulation for the calculation of the entropy change of any cyclical process, reversible or not, i.e.

$$\Delta S_{Clausius}[Cycle] = -\frac{Q_h}{T_h} + \frac{Q_c}{T_c} \qquad (3.16)$$

Substitution in equation (3.16) of the *actual* quantities of heat leaving the hot reservoir and reaching the cold reservoir in a given cyclical process will produce what this formulation deems the correct entropy change for the universe of such a process.

3.4 THE LAW OF INCREASING ENTROPY

Let us now consider the side by side operation of a simple reversible engine and its irreversible counterpart, both operating between the same two heat reservoirs of temperatures T_h and T_c, and both receiving an equal amount of heat Q_h from the hot reservoir. As previously mentioned, the dissipative processes taking place in the irreversible cycle produce a diminished work output for this process, and on reason of this the rejection of a larger amount of heat to the cold reservoir than that rejected in its reversible counterpart. In what follows the irreversible operation will be identified with the sub index '*irr*'.

For the reversible and irreversible operations, in that order, we can write

$$Q_h = Q_c + W \tag{3.17}$$

$$Q_h = Q_{c,irr} + W_{irr} \tag{3.18}$$

The proper combination of the previous equations leads to

$$W - W_{irr} = Q_{c,irr} - Q_c \tag{3.19}$$

Since

$$W \rangle W_{irr} \tag{3.20}$$

Then

$$Q_{c,irr} \rangle Q_c \tag{3.21}$$

Let us now write the following equation for the entropy change of the irreversible cycle

$$\Delta S_{Clausius}[Cycle]_{irr} = -\frac{Q_h}{T_h} + \frac{Q_{c,irr}}{T_c} \tag{3.22}$$

The addition and subtraction of Q_c / T_c to the right hand side of the previous equation leads, after rearrangement, to

$$\Delta S_{Clausius}[Cycle]_{irr} = (-\frac{Q_h}{T_h} + \frac{Q_c}{T_c}) + \frac{(Q_{c,irr} - Q_c)}{T_c} \quad (3.23)$$

In the previous equation we can recognize that the first tem of the right hand side represents the entropy change for the reversible cycle. The contribution of this term, as eq. (3.15) shows, is equal to zero. Therefore

$$\Delta S_{Clausius}[Cycle]_{irr} = \frac{Q_{c,irr} - Q_c}{T_c} \quad (3.24)$$

The combination of equations (3.21) and (3.24) leads us, finally, to

$$\Delta S_{Clausius}[Cycle]_{irr} \rangle 0 \quad (3.25)$$

Thus, while eq. (3.15) shows that the reversible operation takes place at constant total entropy, eq. (3.25) shows, on its part, that the irreversible operation is accompanied by an entropy increase. Results of this sort led Clausius to a combined proposition expressing that "*...the algebraic sum of all the transformations which occur in a cyclical process must always be positive or in the limit equal to zero...*" [Clausius, 1879, p. 213]. Eventually Clausius extended the results of his analysis beyond the confines of cyclical processes:

"*...The theorem which...was enunciated in reference to circular processes only...has thus assumed a more general form, and may be enunciated thus:- The algebraic sum of all the transformations occurring in any alteration of condition whatever can only be positive, or as an extreme case, equal to nothing...*" [Clausius 1862, p. 211]

With this generalization at hand, he was now in the position of stating his results in the following form "*The entropy of the world tends towards a maximum"* [Planck, p. 104], a proposition also known as the law of increasing entropy.

If by W_{lost} we designate the difference in work outputs in the reversible and irreversible operations, i.e. $W_{lost} = W - W_{irr} = Q_{c,irr} - Q_c$, then equation (3.24) can be also be written in the following form

$$\Delta S_{Clausius}[Cycle]_{irr} = \frac{W_{lost}}{T_c} \qquad (3.26)$$

The lost work, it should be emphasized, represents the work that the irreversible operation was incapable of producing on reason of the dissipative effects there operating. As eq. (3.19) shows, this lost work appears as that extra amount of heat the cold reservoir receives above that received in the reversible operation.

This result allows us to get a glimpse at what an entropy increase actually measures: The degradation of work into heat.

3.5 TWO MORE TRANSFORMATIONS

Two more transformations and their values are needed in order to complete the set that defines Clausius' formulation of the second law of thermodynamics. These are the irreversible transformation of heat from high to low temperature: $[Q_c(T_h) \rightarrow Q_c(T_c)]_{irr}$, and the irreversible transformation of work into heat: $[W \rightarrow Q(T)]_{irr}$.

As known, an irreversible heat transfer consists in the direct passage of a given amount of heat from a hot to a cold reservoir. Since the changes produced by this process reduce to those experienced by the reservoirs, the total entropy change for this process will be given by the combination of the entropy change experienced by the hot reservoir of temperature T_h in releasing an amount of heat Q_c, and the one experienced by the cold reservoir of temperature T_c in absorbing that same amount of heat. Through a direct application of the previously given definition for the entropy function, $\Delta S = Q_{rev} / T$ we can determine that those entropy changes amount, respectively, to $-Q_c / T_h$ and Q_c / T_c. Therefore

$$\Delta S_{Clausius}[Q_c(T_h) \rightarrow Q_c(T_c)]_{irr} = -\frac{Q_c}{T_h} + \frac{Q_c}{T_c} = Q_c(\frac{1}{T_c} - \frac{1}{T_h}) \qquad (3.27)$$

The entropy change for $[W \rightarrow Q(T)]_{irr}$ can, on the other hand be obtained in reference to the process represented in Figure 2. There, Figure 2(a) depicts the two transformations produced by one cycle in the operation of a reversible cyclical process. Their entropy changes are those given by equations (3.9) and (3.11). Let us now take the work generated by the cycle, and as shown in Figure 2(b), irreversibly degrade it via a frictional process into an equivalent amount of heat Q that ends up in the cold reservoir. This degradation has the effect of transforming the original reversible universe of the cyclical process shown in 2(a), into the irreversible universe of the direct heat transfers shown in 2(c). The reason for this change in the nature of these transformations can be traced to the degradation of W. In 2(a), with W available, all we have to do to send both Q and Q_c back to the hot reservoir is to use W to propel, to run the inverse cycle. In 2(c) W is no longer available. Any transfer of Q and Q_c back to the hot reservoir will of necessity leave a permanent change in that body called to supply the missing work. This is the reason for the '*irr*' qualifier accompanying the transformations in 2(c).

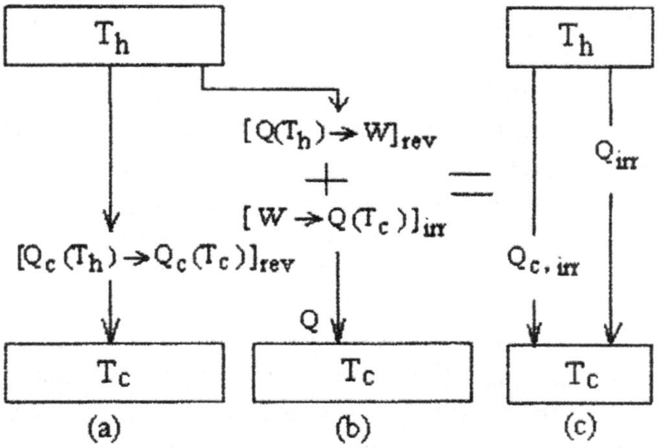

Figure 2. The combination of the transformations associated to the reversible cyclical process shown in (a) with the irreversible, frictional degradation of the work W into heat of temperature T_c shown in (b), reduces to the two irreversible transfers of heat shown in (c), where Q_{irr} and $Q_{c,irr}$ stand, respectively, for $[Q(T_h) \to Q(T_c)]_{irr}$ and $[Q_c(T_h) \to Q_c(T_c)]_{irr}$.

It should also be noted that in 2(c), $[Q(T_h) \to Q(T_c)]_{irr}$ is shorthand for the concatenation heat-to-work/work-to-heat involving Q, i.e.

$$[Q(T_h) \to W]_{rev} + [W \to Q(T_c)]_{irr} = [Q(T_h) \to Q(T_c)]_{irr} \qquad (3.28)$$

According to equation (3.27), the following expression can be written for $[Q(T_h) \to Q(T_c)]_{irr}$

$$\Delta S[Q(T_h) \to Q(T_c)]_{irr} = -(Q / T_h) + (Q / T_c) \qquad (3.29)$$

The combination of the two previous equations produces

$$[Q(T_h) \to W]_{rev} + [W \to Q(T_c)]_{irr} = -(Q / T_h) + (Q / T_c) \qquad (3.30)$$

Further combination of the previous equation with eq. (3.9) produces the following result

$$\Delta S_{Clausius}[W \rightarrow Q(T_c)]_{irr} = \frac{Q}{T_c} \qquad (3.31)$$

Clausius Formulation of the second law of thermodynamics stands characterized in Table 1 by the values or entropy changes it assigns to the different possible transformations of heat into work, heat from one temperature to another, or vice versa.

Even if it is a common belief that what is being taught nowadays as second law thermodynamics is centered on Clausius' analysis, the truth of the matter is that current second law thermodynamics is a radically different version of Clausius' thermodynamics as represented by the set of values contained in Table 1.

Before this matter can be further discussed, we will subject Clausius' results to a logical analysis which will show that Clausius paramount result, the law of increasing entropy, is nothing more than a logical fallacy, i.e. a conclusion contradicted by its very premises.

Transformation	Clausius
$[Q(T_h) \rightarrow W]_{rev}$	Eq. (3.9): $-\frac{Q}{T_h}$
$[W \rightarrow Q(T_h)]_{rev}$	Eq. (3.10): $\frac{Q}{T_h}$
$[W \rightarrow Q(T_c)]_{irr}$	Eq. (3.31): $\frac{Q}{T_c}$
$[Q_c(T_h) \rightarrow Q_c(T_c)]_{rev}$	Eq. (3.11): $Q_c(\frac{1}{T_c} - \frac{1}{T_h})$
$[Q_c(T_c) \rightarrow Q_c(T_h)]_{rev}$	Eq. (3.12): $Q_c(\frac{1}{T_h} - \frac{1}{T_c})$
$[Q_c(T_h) \rightarrow Q_c(T_c)]_{irr}$	Eq. (3.27): $Q_c(\frac{1}{T_c} - \frac{1}{T_h})$

Table 1. Clausius values or entropy changes for the transformations taking place in reversible engines and refrigerators. The table includes the values for the irreversible transformation of work into heat, and heat from a higher to a lower temperature.

CHAPTER 4: THE FLAWS IN CLAUSIUS FORMULATION

It has to be mentioned that at no place along his work Clausius identified the process or processes behind the two transformations that one cycle in the operation of a reversible cyclical process can be reduced to. In other words, even if these transformations constitute an essential part of his work on the second law, no information was by him forwarded as to the path followed by this heat being transferred from the hot to the cold reservoir, nor about the path leading to the work output of the cycle. This bit of knowledge is essential if a criticism of Clausius results can be advanced. The argument leading to the dissection of a simple reversible cyclical process into its two constitutive transformations is presented below. Once identified, these transformations will be used to conduct a check on Clausius' values presented in Table 1, as well as to identify the logically unsustainable position Clausius adopts in using the combinations of the values of these two transformations to produce his law of increasing entropy.

4.1 THE NATURE OF THE TRANSFORMATIONS

As previously stated in reference to the reversible engine represented in Figure 1, the amount of heat taken in by the ideal gas from the hot reservoir along *AB* is larger than that it gives out to the cold reservoir along *CD*. This fact allows us to realize that there exists some point *E* on isotherm AB such that in its transit along *EB* the ideal gas takes from the hot reservoir an amount of heat Q_c identical to that given out to the cold reservoir along *CD*. The approximate location of point *E* is shown in Figure 3.

Assuming now that the operation of the cycle is to start at point *E*, consideration will be given to the effect of the concatenation of processes *EB-BC-CD-DA*. Actually, on reason of the fact that the changes brought about by adiabatic and reversible process *BC* are precisely cancelled by adiabatic and reversible process *DA*, the effect of the said concatenation reduces to that produced by the combination of processes *EB* and *CD*.

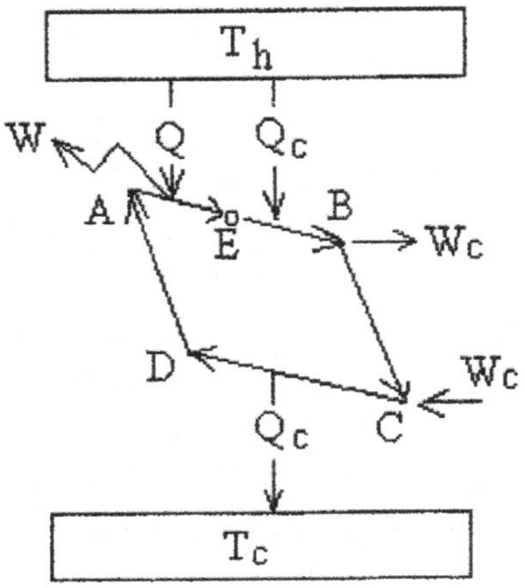

Figure 3. Point E has been selected in a way such that in its expansion from E to B the variable body transforms an amount of heat Q_c absorbed from the hot reservoir, into an equivalent amount of work W_c. This work, expended in carrying on compression CD, ends up in the cold reservoir as the equivalent amount of heat Q_c. The previous consideration, alongside the fact that the effects of process BC are precisely cancelled by those of process DA, allows us to realize that the sole effect of concatenation EB-BC-CD-DA is that of reversibly transferring the amount of heat Q_c from the hot to the cold reservoir. The process bringing the cycle to its conclusion, isothermal and reversible expansion AE, transforms, on its part, the amount of heat Q into the equivalent amount of work W.

Along *EB* the hot reservoir transfers to the ideal gas an amount of heat Q_c. The ideal gas, on its part, transforms this heat into an equivalent amount of work W_c. Along *CD*, on the other hand, the amount of work W_c expended in carrying on this isothermal and reversible compression finds its way, as an equivalent amount of heat Q_c, into the cold reservoir. This $[Q_c(T_h) \rightarrow W_c]/[W_c \rightarrow Q_c(T_c)]$ combination produced by processes *EB* and *CD* allows us to realize that what this concatenation manages to effect is the transfer of Q_c from the hot to the cold reservoir. In other words, it is the concatenation of processes *EB-BC-CD-DA* the one responsible for the reversible transformation of Q_c from T_h to T_c which, as already stated, is represented as $[Q_c(T_h) \rightarrow Q_c(T_c)]_{rev}$.

Let us now recognize that the process *AE* needed to bring the cycle to its conclusion is an isothermal and reversible expansion at the temperature of the hot reservoir. Through it the ideal gas absorbs an amount of heat Q from the hot reservoir and transforms it into an equivalent amount of work W. Since no other effect but this can be associated to this expansion, it follows that it is *AE* the one responsible for the net work output of the cycle, or in other words, for bringing forward the transformation of the amount of heat Q of temperature T_h into an equivalent amount of work , or symbolically $[Q(T_h) \rightarrow W]_{rev}$. Again, the *rev* sub-index included in the notation for this and the previous transformation indicates that both of these transformations, as components of a reversible process, are reversible themselves.

4.2 THE LOGICAL FLAW OF CLAUSIUS' THERMODYNAMICS

Let us center our attention on the process responsible for the work output of the cycle, i.e. on isothermal and reversible expansion *AE* in Figure 3. For effects of clarity this process, responsible for transformation $[Q(T_h) \rightarrow W]_{rev}$, will be represented as shown in Figure 4(a). As previously explained, here the ideal gas acting as variable body transforms quantitatively into work (W) an amount of heat (Q) absorbed from the hot reservoir. The very first thing that has to be understood in regard to this process is that work production is concomitant to the transfer of heat from the hot reservoir to the gas. As soon as an element of heat dQ passes from the reservoir to the gas it is transformed into the equivalent amount of work dW. If this isn't so then the process, contrary to our assumption, can not be isothermal.

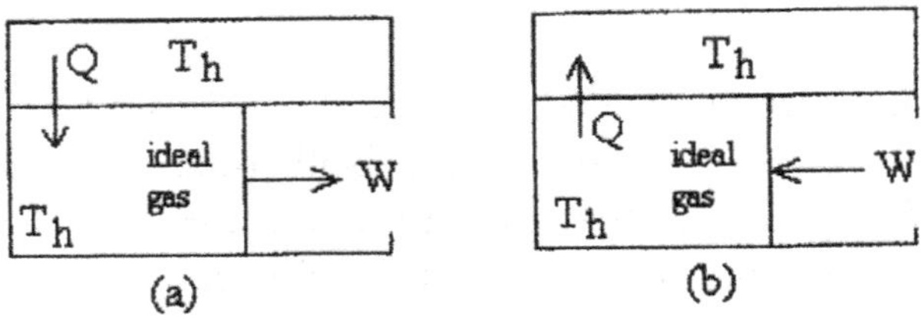

Figure 4. Upon completion, the isothermal and reversible expansion *AE* represented in (a) manages to transform an amount of heat Q of temperature T_h into an equivalent amount of work W. Every element of work dW produced as the expansion takes place demands, however, the flow of the equivalent amount of heat dQ from the heat reservoir to the ideal gas. This means that there are two transformations taking place as the reversible expansion proceeds: The transformation of heat from the reservoir to the gas, and the transformation of this heat into work. The isothermal and reversible compression shown in (b) represents, on its part, process *EA*, the inverse of process *AE*. Note that each and every change produced by *AE* is reversed by *EA* and because of this the coupling of (a) and (b) take the universe back to its original condition.

This argument leads us to the realization that there are actually two transformations taking place in the said process: The reversible transference of heat from the reservoir to the ideal gas, and the necessarily reversible transformation of this heat into work by the reversible expansion of the gas.

In reference to the finite amount of heat Q, and its equivalent amount of work W, the situation previously described leads to the following expression for the entropy change of such a reversible process:

$$\Delta S_{Clausius}[AE]_{rev} = \Delta S[Q(T_h) \to Q(T_h)]_{rev} + \Delta S[Q(T_h) \to W]_{rev} =$$

$$0 + \Delta S[Q(T_h) \to W]_{rev} \qquad (4.1)$$

The zero entropy change substituted in equation (4.1) for the heat transfer taking place in process 4(a) is a consequence of the fact indicated by equation (3.13) that any transference of heat between two bodies at essentially the same temperature takes place at constant total entropy. Being this so, the entropy change for process 4(a), as indicated by the rightmost section of equation (4.1), will be solely and uniquely determined by the entropy change for the heat to work transformation there taking place. All we need to do to now is to replace $\Delta S[Q(T_h) \to W]_{rev}$ with its actual value. No problem here. All we have to do is to look back at Table 1. There we find that the value in question amounts to $-Q/T_h$. Therefore the total entropy change for the heat-to-work transformation taking place in a reversible cyclical process, written now in transformation notation, amounts to

$$\Delta S_{Clausius}[Q(T_h) \to W]_{rev} = -Q/T_h \qquad (4.2)$$

It has to be clearly understood that the entropy change given by equation (4.2) corresponds to the *universe* of the isothermal and reversible expansion *EA* or equivalently, to its *total* entropy change. As such it subsumes all the entropy changes involved in *AE* or, equivalently, in transformation $[Q(T_h) \to W]_{rev}$; these being the $-Q/T_h$ entropy change sustained by the hot reservoir in releasing the amount of heat Q; the Q/T_h entropy change experienced by the ideal gas in absorbing the amount of heat Q, and finally the $-Q/T_h$ entropy change associated to the actual transformation of heat into work through the isothermal and reversible expansion of the ideal gas. In other words, in Clausius thermodynamics the universe of this transformation is actually negentropic, meaning that a net total entropy decrease proportional to the amount of work produced has taken place. For the isothermal and reversible compression shown in Figure 4(b) with which we have represented process *EA* or equivalently, transformation $[W \to Q(T_h)]_{rev}$, the following expression for its entropy change now can be written

$$\Delta S_{Clausius}[EA]_{rev} = \Delta S[W \to Q(T_h)]_{rev} + \Delta S[Q(T_h) \to Q(T_h)]_{rev} =$$

$$0 + \Delta S[W \to Q(T_h)]_{rev} \tag{4.3}$$

Substitution of the appropriate Clausius's value in equation (4.3) leads to the following entropy change for transformation $[W \to Q(T_h)]_{rev}$

$$\Delta S_{Clausius}[W \to Q(T_h)]_{rev} = Q / T_h \tag{4.4}$$

The same considerations previously advanced in regard to equation (4.2) can be called to identify the entropy change given by equation (4.4) as the entropy change for the universe of isothermal and reversible compression *EA* or equivalently, of transformation $[W \to Q(T_h)]_{rev}$ As equation (4.4) expresses, the reversible universe under consideration is entropic in the amount Q / T_h.

As can be inferred from the previous paragraphs the terms universe entropy change and total entropy change signify the very same thing, this is, the combined values of all the entropy changes associated to the occurrence of the process being considered.

Let us now determine the total entropy change for the reversible transformation of Q_c from the hot to the cold reservoir taking place through the concatenation of processes *EB-BC-CD-DA*, as depicted in Figure 3. As should be recalled, in it process *EB* represents an isothermal and reversible expansion at the temperature of the hot reservoir. Through it an amount of work W_c is produced from an equivalent amount of heat Q_c absorbed by the ideal gas from the hot reservoir. Process *CD* represents, on its part, an isothermal and reversible compression at the temperature of the cold reservoir. Through it the amount of work W_c expended in carrying on this compression ends up as an equivalent amount of heat in the cold reservoir. For the reason already given in regard to processes *BC* and *DA* we'll have that the effects of the concatenation reduces to those of the work producing isothermal and reversible expansion *EB*, and work consuming isothermal and reversible compression *CD*. As should be obvious to the reader, the analysis of these processes is to proceed along the line used for the analysis of processes 4(a) and 4(b) above performed. The entropy change for process *EB* is thus determined by the summation of the entropy changes of the two transformations there taking place: the transfer of Q_c from the hot reservoir to the gas, and the transformation of this heat into an equivalent amount of work W_c via the expansion of the gas, i.e.

$$\Delta S_{Clausius}[EB]_{rev} = \Delta S[Q_c(T_h) \rightarrow Q_c(T_h)]_{rev} + \Delta S[Q_c(T_h) \rightarrow W_c]_{rev} =$$

$$-Q_c / T_h \tag{4.5}$$

The entropy change for process *CD* is, on its part, determined by the summation of the entropy changes associated to the transformation of W_c into an equivalent amount of heat via the compression of the gas, and that associated to the transfer of this heat from the gas to the cold reservoir, i.e.

$$\Delta S_{Clausius}[CD]_{rev} = \Delta S[W_c \rightarrow Q_c(T_c)]_{rev} + \Delta S[Q_c(T_c) \rightarrow Q_c(T_c)]_{rev} =$$
$$Q_c / T_c \tag{4.6}$$

The entropy change for the transformation $[Q_c(T_h) \rightarrow Q_c(T_c)]_{rev}$ can now be written as the combination of the entropy changes for processes *EB* and *CD* as follows

$$\Delta S_{Clausius}[Q_c(T_h) \rightarrow Q_c(T_c)]_{rev} = Q_c(\frac{1}{T_c} - \frac{1}{T_h}) \quad (4.7)$$

The fact that the entropy change shown in equation (4.7) has been obtained via the summation of all the entropy changes subsumed by the concatenation defining the reversible transfer of $[Q_c(T_h) \rightarrow Q_c(T_c)]_{rev}$ from the hot to the cold reservoir makes it the total entropy change for this transformation. The fact that by original assumption $T_h \rangle T_c$ allows us to re-express this by saying that the entropy change for the universe of reversible transformation $[Q_c(T_h) \rightarrow Q_c(T_c)]_{rev}$ is positive and of magnitude $Q_c(T_h - T_c)/(T_h T_c)$.

Suppose now that once the ideal gas has reached condition *A*, as represented in Figure 3, we make it retrace its steps, i.e., to travel the *ADCBE* path. In this situation, process *DC* represents an isothermal and reversible expansion at the temperature of the cold reservoir. Through it an amount of heat Q_c absorbed by the gas from the cold reservoir is transformed into an equivalent amount of work W_c. Process *BE*, on its part, represents an isothermal an reversible compression at the temperature of the hot reservoir. In it the amount of work W_c here expended ends up as an equivalent amount of heat in the hot reservoir. The combination of the effects of this heat-to-work/work-to-heat combination produces as its net result the reversible transfer of the amount of heat Q_c from the cold to the hot reservoir, i.e. $[Q_c(T_c) \rightarrow Q_c(T_h)]_{rev}$

For the same reason advanced above, the analysis that follows will be restricted to the changes associated to processes DC and BE, whose entropy changes will be written without further explanation as follows

$$\Delta S_{Clausius}[DC]_{rev} = \Delta S[Q_c(T_c) \to Q_c(T_c)]_{rev} + \Delta S[Q_c(T_c) \to W_c]_{rev} =$$

$$-Q_c / T_c \tag{4.8}$$

$$\Delta S_{Clausius}[BE]_{rev} = \Delta S[W_c \to Q_c(T_h)]_{rev} + \Delta S[Q_c(T_h) \to Q_c(T_h)]_{rev} =$$

$$Q_c / T_h \tag{4.9}$$

The total entropy change for this transformation can now be written as the summation of the two entropy changes given above

$$\Delta S_{Clausius}[Q_c(T_c) \to Q_c(T_h)]_{rev} = Q_c(\frac{1}{T_h} - \frac{1}{T_c}) \tag{4.10}$$

As before done in regard to equations (4.2), (4.4) and (4.7), and for the same reasons, it will be said that the entropy change for the universe of the reversible transfer of the amount of heat Q_c from the cold to the hot reservoir, i.e. for the universe of transformation $[Q_c(T_c) \to Q_c(T_h)]_{rev}$, amounts to $Q_c(T_c - T_h)/(T_h T_c)$, a negative quantity.

The previous analysis has allowed us to introduce an element of precision in the argument concerning Clausius' values in the sense that now we can pair or associate each of Clausius values or entropy changes, $-Q/T_h$ and $Q_c(\frac{1}{T_c} - \frac{1}{T_h})$, with the universe of processes AE and $EBCDA$, as represented in Figure 3. Reiterating, that those entropy changes correspond to the universe of their associated process or processes comes from the fact that in their calculation each and every entropy change arising in each of those processes has been taken into account.

At the light of these results the logical aberration permeating Clausius' thermodynamics can now be identified. It takes form in the contradiction existing between the *non-zero entropy changes* he assigns to the universes of his reversible transformations, with his other claim contained in the law of increasing entropy stating that the universe of every reversible process is accompanied by a *zero entropy change*. The irony of it all is that it was by combining the *non-zero* entropy changes for the universes of these transformations that Clausius managed to produce –as shown by equation (3.14)- a *zero* entropy change for the universe of the reversible cyclical processes and through it, eventually claim the same result for all reversible processes "*...The theorem which...was enunciated in reference to circular processes only...has thus assumed a more general form, and may be enunciated thus:- The algebraic sum of all the transformations occurring in any alteration of condition whatever can only be positive, or as an extreme case, equal to nothing...*" [Clausius, 1862, p. 211]. The fact that Clausius' conclusion regarding the zero entropy change for the universe of reversible processes is actually negated by the very premises –the non-zero entropy changes for the universes of the reversible transformations- on which such a conclusion finds sustenance, makes Clausius' analysis a good example of bad science.

A similar conclusion can without difficulty be obtained in regard to the law of increasing entropy according to which while reversible processes take place at constant total entropy; a positive entropy change accompanies the universe of those irreversible. The essential message of this law can be expressed by saying that entropy production is an exclusive trait of irreversible processes. A quick look at Clausius' values demonstrates however, that this is not the case. There we found two reversible processes: $[W \rightarrow Q(T_h)]_{rev}$ and $[Q_c(T_h) \rightarrow Q_c(T_c)]_{rev}$, whose occurrence is respectively accompanied by positive total entropy changes in the amounts of Q/T_h and $Q_c(\frac{1}{T_c}-\frac{1}{T_h})$.

Even more, in Clausius' second law processes with a total negative entropy change are impossible –forbidden- as they supposedly can be likened or reduced to the spontaneous flow of heat from a colder to a hotter body. Not so! The total entropy change for the (necessarily) reversible transformation of heat into work through an isothermal and reversible expansion is actually negative; yet this process has no connection, nor parentage, nor affiliation with heat spontaneously flowing from a colder to a hotter body. An isothermal and reversible expansion is an allowed –not forbidden- process yet, contrary to Clausius thermodynamics' wisdom, it proceeds with a net total entropy decrease!

If we compare a reversible process with a duck and an irreversible process with a swan, and at the same time represent the constant, and increasing total entropy qualities with the colors black and white, respectively, then what the law of increasing entropy states is that in Clausius' zoo only two animals can be found: black ducks and white swans. This however, happens not to be the case. A quick perusal of the zoo grounds allowed us to find that coexisting with the above birds there were not only white ducks (total entropy increasing reversible processes), but also a variety whose existence is not even acknowledged by his law of increasing entropy: red ducks (total entropy decreasing reversible process). The menagerie was actually more diverse than Clausius' law of increasing entropy led us to believe.

At the light of the logical shortcomings of Clausius' thermodynamics no other course of action can be taken but that consisting in a total overhaul that will, of necessity, start with a re-evaluation of the values of the transformations, and from there on to whatever general statement about the total entropy change for the universe of irreversible processes this revision leads to. The discussion offered in the next chapter identifies current second law thermodynamics as a –however veiled- correction to Clausius' thermodynamics.

CHAPTER 5: CURRENT SECOND LAW THERMODYNAMICS

5.1 CLAUSIUS' FORMULATION VERSUS CURRENT THERMODYNAMICS, OR THE CONTRADICTION BETWEEN TWO FUNDAMENTALLY DIFFERENT VERSIONS OF THE SECOND LAW

The law that entropy always increases holds, I think, the supreme position among the laws of Nature. If someone points out to you that your pet theory of the universe is in disagreement with Maxwell's equations – then so much the worse for Maxwell's equations. If it is found to be contradicted by observation –well, these experimentalists do bungle things sometimes. But if your theory is found to be against the second law of thermodynamics I can give you no hope; there is nothing for it but to collapse in deepest humiliation.
Sir Arthur Stanley Eddington, The Nature of the physical World, 1915

Roma locuta est; causa finita est.
Rome has spoken; the case is concluded.
St. Augustine,
http://en.wikiquote.org/wiki/Augustine_of_Hippo

Most students and studious of thermodynamics believe that second law thermodynamics as thought nowadays, and as have been taught for a long number of years, is simply a compendium of Clausius' work; a little polished perhaps; a lot extended, for certain, but all in all Clausius' work. This belief, however, bears no correspondence with reality. The version of second law thermodynamics that the present author, and you the reader, have been exposed to is actually a radically different version of the second law than that Clausius struggle to construct. From Clausius, current second law thermodynamics (CT) for the most part retained the definition of the entropy function, the equation quantifying the entropy changes of cyclical processes, as well as the law of increasing entropy stating that the total entropy change for a reversible process is zero and positive for all other. The rest of Clausius' construction, it might be said, was disregarded. As to the reason behind this move this author firmly believes that at some point or another, the previously described flaws in Clausius' analysis might have surfaced. At that point the law of increasing entropy –the supreme law of nature, according to Eddington- might have already taken central stage in thermodynamics, and most likely in philosophy / politics / economics / demography / religion etc. It is very likely that at this point a conscious choice was made –By whom? I do not know!- to modify Clausius' values in a way such that the new formulation emerging, besides of being devoid of Clausius analysis' flaws, was also self-consistent and in complete agreement with the law of increasing entropy. No effort was made to introduce the necessary corrections into Clausius' analysis in order to fulfill the goal he unsuccessfully pursued; that of bringing forward a law which reflected nature's behavior in regard to heat-work inter-conversions. In modifying Clausius' values to retain or guarantee the validity of the law of increasing entropy what we got in return was a distorted image of reality, or to be more precise, an image that only under certain circumstances corresponds to that of nature; or even perhaps, a construction to suit the purposes of an ideology. This last explanation becomes even more viable once some of the contenders cheering for, or against the law of increasing entropy are identified, as this quote of Bazarov allows:

The (second) law has caught the attention of poets and philosophers and has been called the greatest achievement of the nineteenth century. Engels disliked, for it supported opposition to Dialectical Materialism, while Pope Pius XII regarded it as proving the existence of a higher being.
Ivan Bazarov, Thermodynamics (1964)
Wikiquote, http://en.wikiquote.org/wiki/Thermodynamics

To identify the point at which current thermodynamics separates from Clausius' analysis all we have to do is to take another look a that simple process called an isothermal and reversible expansion of an ideal gas represented in Figure 4(a), the same that embodies the reversible transformation of heat into work in a reversible cyclical process, and to which Clausius' perspective, as described in section 4.2, assigns an entropy change in the amount of $-Q/T_h$.

Irrespective of which perspective, Clausius' or current thermodynamics, is adopted to study the isothermal and reversible expansion, one fact remains true, before any element of work dW can be delivered by the ideal gas to the mechanical reservoir, the equivalent amount of heat dQ has first to flow from the heat reservoir to the gas. The combination of all these infinitesimal transformations $[dQ(T_h) \rightarrow dQ(T_h)]_{rev}$ and $[dQ(T_h) \rightarrow dW]_{rev}$, will eventually produce the finite changes to which the following equation refers

$$\Delta S_{CT}[4(a)]_{rev} = \Delta S[Q(T_h) \rightarrow Q(T_h)]_{rev} + \Delta S[Q(T_h) \rightarrow W]_{rev} =$$

$$0 + \Delta S[Q(T_h) \rightarrow W]_{rev} \tag{5.1}$$

In the previous equation the sub-index CT stands for 'current thermodynamics'. It should be noted that equation (5.1) is identical to equation (4.1). The zero appearing in both of them stems from the fact that also independent of the perspective adopted is the fact that no entropy change is produced by the transference of a given amount of heat between two bodies of the same temperature. At difference of equation (4.1), equation (5.1) will not be solved for $\Delta S_{CT}[4(a)]_{rev}$. The reason being that unless you read line by line the Physical Chemistry of G. M. Barrow, or the Thermodynamics of Lewis and Randall, as reviewed by Pitzer and Brewer, you will never find a statement about the entropy change current thermodynamics ascribes to transformation $[Q(T_h) \rightarrow W]_{rev}$. We know, however, that the value it assigns to the total entropy change of process 4(a) is zero [Klotz and Rosenberg, p. 133; Maron and Prutton, p.

170]. Substitution of $\Delta S_{CT}[4(a)]_{rev} = 0$ in eq. (5.1) leads us to the value current thermodynamics assigns to the (necessarily) reversible transformation of heat into work:

$$\Delta S_{CT}[Q(T_h) \to W]_{rev} = 0 \qquad (5.2)$$

Thus, while for Clausius the transformation of heat into work is a negentropic process i.e. a process with a total entropy change of $-Q/T$, for current thermodynamics such a process proceeds at constant total entropy.

Current thermodynamics cares not to explain the rationale behind its position. It however acknowledges it as the following quotes attest

...We are at liberty, you should recognize, to ascribe any features to this new entropy function that we like, the requirement being that we construct a function that is self-consistent and allows us to form a useful expression for the second law. In this vein we further specify that for all processes: $dS_{mech\ res} = 0$

[G. M. Barrow, 1973, p. 168]

...If a heat engine operates reversibly and passes through a whole number of complete cycles...it will itself suffer no change of entropy. Hence all the entropy changes are in the rest of the system, and these must sum up to zero in a reversible process...If q_h *is the heat taken from the hot reservoir at* T_h *and* q_c *is the heat given to a cold reservoir at* T_c*, then the increase of entropy of the hot reservoir is* $-q_h/T_h$*, and that of the cold reservoir* q_c/T_c *(then)*

$$-\frac{q_h}{T_h} + \frac{q_c}{T_c} = 0$$

...The work term does not appear in (this) equation, since it involves no entropy...

[Pitzer and Brewer (Revision of Lewis and Randall) 1961, p. 95]

In Barrow's quote the statement that $dS_{mech\ res} = 0$ (with '*mech res*' referring to the mechanical reservoir) is meant to signify that a zero total entropy change accompanies the transformation of heat into work.

The fact that the inverse of a given reversible process produces the precise opposite effects of the latter means that the total entropy changes for one and the other will have the same magnitude but opposite sign. At the light of this consideration it follows that

$$\Delta S_{CT}[W \rightarrow Q(T_h)]_{rev} = 0 \tag{5.3}$$

The results expressed by equations (5.2) and (5.3) are a simple consequence of the fact that for current thermodynamics the total entropy change for any isothermal and reversible ideal gas expansion or compression –the processes responsible for the reversible inter-conversions of heat and work as they take place in reversible cyclical processes- is equal to zero.

Armed with this knowledge we will proceed to evaluate the total entropy change assigned by current thermodynamics to the reversible transformation of heat from one temperature to another as it occurs in a reversible cyclical process. With reference to Figure 3 and keeping in mind that for the reasons already advanced in regard to processes *BC* and *DA*, the entropy change for this transformation is determined by the entropy changes of isothermal and reversible processes *EB* and *CD*, we will have that

$$\Delta S_{CT}[Q_c(T_h) \rightarrow Q_c(T_c)]_{rev} = \Delta S[Q_c(T_h) \rightarrow W_c]_{rev} + \Delta S[W_c \rightarrow Q_c(T_c)]_{rev} =$$

$$0 + 0 = 0 \tag{5.4}$$

And just like for $[Q(T_h) \rightarrow W]_{rev}$, the transfer of heat taking place in a reversible cyclical process also has, for current thermodynamics, an entropy change of zero.

The same reasoning advocated in the transit of equation (5.2) to (5.3) allows us to write the following equation for the reversible transfer of Q_c from the cold to the hot reservoir

$$\Delta S_{CT}[Q_c(T_c) \rightarrow Q_c(T_h)]_{rev} = 0 \qquad (5.5)$$

According then to the values shown by equations (5.2) and (5.4), for current thermodynamics, as it was proved before for Clausius' formulation, the total entropy change for a reversible cyclical process amounts to zero, i.e.

$$\Delta S_{CT}[Cycle]_{rev} = \Delta S_{CT}[Q(T_h) \rightarrow W]_{rev} + \Delta S[Q_c(T_h) \rightarrow Q_c(T_c)]_{rev} =$$

$$0 + 0 = 0 \qquad (5.6)$$

Note here that both Clausius as well as current thermodynamics produce the same result for this process. The path leading to such a result is however, radically different in one and the other. Clausius produces a zero for the total entropy change of the reversible cycle through the addition –as equation (3.14) shows- of two non-zero values; Current Thermodynamics does it, on its part, through the addition –as the previous equation shows- of two zeros.

In order to be able to fully contrast Clausius' formulation with current thermodynamics we must develop for the latter a Table of values similar to Table 1. In order to do this we must ascertain the values current thermodynamics ascribes to the following transformations: $[W \rightarrow Q(T)]_{irr}$, and $[Q_c(T_h) \rightarrow Q_c(T_c)]_{irr}$.

The value of the former transformation can be found in the Thermodynamics of Lewis and Randall

...If an amount of work δW is degraded to heat of temperature T, the increase in entropy is $dS_{irr} = \delta W / T$ ...
[Pitzer and Brewer (Revision of Lewis and Randall) 1961, pp. 85-86]

Using our notation, this value will be represented as follows

$$\Delta S_{CT}[W \rightarrow Q(T)]_{irr} = Q / T \qquad (5.7)$$

The fact that the entropy change for transformation $[Q_c(T_h) \to Q_c(T_c)]_{irr.}$ originates in a simple application of the defining equation of the entropy function to both of the bodies involved in such a transfer, makes the entropy change for this transformation identical in both formulations, i.e.

$$\Delta S_{CT}[Q_c(T_h) \to Q_c(T_c)]_{irr} = -\frac{Q_c}{T_h} + \frac{Q_c}{T_c} = Q_c(\frac{1}{T_c} - \frac{1}{T_h}) \tag{5.8}$$

In order to make easier the contrast between Clausius' version of the second law and that furnished by current second law thermodynamics, the values for the transformations in both of these formulations are presented vis-à-vis in table 2.

Transformation	Clausius	CT
$[Q(T_h) \to W]_{rev}$	$-\frac{Q}{T_h}$	Eq. (5.2): 0
$[W \to Q(T_h)]_{rev}$	$\frac{Q}{T_h}$	Eq. (5.3): 0
$[W \to Q(T_h)]_{irr}$	$\frac{Q}{T_h}$	Eq. (5.7) $\frac{Q}{T_h}$
$[Q_c(T_h) \to Q_c(T_c)]_{rev}$	$Q_c(\frac{1}{T_c} - \frac{1}{T_h})$	Eq. (5.4): 0
$[Q_c(T_c) \to Q_c(T_h)]_{rev}$	$Q_c(\frac{1}{T_h} - \frac{1}{T_c})$	Eq. (5.5): 0
$[Q_c(T_h) \to Q_c(T_c)]_{irr}$	$Q_c(\frac{1}{T_c} - \frac{1}{T_h})$	Eq. (3.27) $Q_c(\frac{1}{T_c} - \frac{1}{T_h})$

Table 2. Clausius set of values for the transformations are shown vis-à-vis those sanctioned by current second thermodynamics (CT)

Notwithstanding the differences made evident by Table 2, the same general equation for the entropy change of cyclical processes, reversible or not, rules in both formulations:

$$\Delta S_{CT}[Cycle] = -\frac{Q_h}{T_h} + \frac{Q_c}{T_c} \tag{5.9}$$

The previous equation, identical to zero in the case of reversible cycles, will necessarily produce positive results in the case of those evolving irreversibly. The fact that any irreversibility in those cycles can be reduced to a direct, irreversible transfer of heat from the hot to the cold reservoir, combined with the fact that any such transfer produces a positive entropy change, leads to the inescapable conclusion that any such cycle will produce an entropy increase. This is the law of increasing entropy in the form it takes for cyclical processes: Constant entropy for reversible cycles, entropy production for those irreversible.

As previously mentioned, it is this author's belief that current second law thermodynamics is the result of a conscious effort to save the law of increasing entropy. It has already been shown that in Clausius thermodynamics the values for the transformations associated to a reversible cyclical process, the ultimate foundation for his law of increasing entropy, are actually in opposition with such a law. As any commonsensical approach to this situation would indicate, such a discrepancy can have three sources:

1. The values for the transformations are wrong but the law of increasing entropy is correct.
2. The values for the transformations are correct and the law of increasing entropy is wrong.
3. Both, the values for the transformations as well as the law of increasing entropy are wrong.

Of these three, the first one was chosen. The reason of this particular choice is, of course unknown, as unknown are the answers to the following questions: Were the other two options seriously studied? If so, who did the research and where were the results of this exploration published, or presented? The answers to these questions, if they exist at all, are unfamiliar to this author. What is certainly possible is that whoever performed these changes was certainly aware that a very simple procedure existed to accomplish the stated goal of salvaging the law of increasing entropy through the simple fact of sweeping Clausius' flaws under the rug. The results of this simple procedure can be seen in table 2. Through the expedient of making zero the entropy change of any reversible transformation and positive the entropy changes of those irreversible, the flaws were no more. Now the law of increasing entropy –obtained by Clausius as a generalization of the zero entropy change for a reversible cyclical process- agrees with the values of both of the transformations involved, as both of them now carry zero entropy changes. The entropy change of any reversible process, constituted solely by reversible transformations had, of necessity, to be zero. Likewise, any irreversible transformation appearing in a process would render the total entropy change positive. Sense in Second Law Thermodynamics was back, or so they thought. The disagreements were solved through the expedient of shooting down those pesky white and red ducks in Clausius' backyard. Only black ducks and white swans remained…or so they thought. In the process of extermination, a rare animal managed to creep into the zoo without the exterminators noticing. This *rara avis,* a black swan, has managed to remain hidden, undetected from the sight of those familiar with the zoo grounds, until now. In the second critique of the coming section we will bring to light the black swan of current second law thermodynamics in the form of a constant entropy irreversible process.

CHAPTER 6: THE FLAWS IN CURRENT SECOND LAW THERMODYNAMICS

6.1 INTRODUCTION

It has been mentioned that out of Clausius analysis, current thermodynamics essentially kept the definition of the entropy function, the equation quantifying the entropy changes of cyclical processes, as well as the law of increasing entropy. As known, this law subsumes two notions: The constant total entropy reversibility criterion, and the total entropy increasing quality of irreversible processes; as well as one corollary: the impossibility of processes with a negative total entropy change.

The following argument will prove, however, that within the rules and definitions of current second law thermodynamics it is possible to produce a counterexample to the two notions and the corollary just mentioned. This counterexample, embodied in a constant total entropy irreversible process, constitutes irrefutable proof of the incorrectness of the law of increasing entropy -the supreme law of nature according to Eddington- setting this way the scene for the development of the definitive version of the second law of thermodynamics.

Let us note, *en passant*, that current thermodynamics sanctions an alternative reversibility criterion as equivalent to the constant total entropy above mentioned. This alternative criterion –transcribed in the second quote below- is due to Planck, and states that it is impossible to recuperate the precise original condition of the universe once an irreversible process has taken place. The counterexample above referred will also prove the non-equivalence between these two criterions.

6.2 THE COUNTEREXAMPLE

6.2.1 An irreversible heat transfer

...The second law of thermodynamics states that there exists in nature for each system of bodies a quantity, which by all changes of the system either remains constant (in reversible processes) or increases in value (in irreversible processes). This quantity is called, following Clausius, the entropy...
[Planck, p. 88]

The significance of the second law of thermodynamics depends on the fact that it supplies a necessary and far reaching criterion as to whether a definite process which occurs in nature is reversible or irreversible...A process which can in no way be completely reversed is termed irreversible, all other processes reversible. That a process may be irreversible, it is not sufficient that it cannot be directly reversed....The full requirement is, that it be impossible, even with the assistance of all agents in nature, to restore everywhere the exact initial state when the process has once taken place. [Planck, pp. 86-87, 84]

The argument that follows, the matter of a previous publication by the author [Íñiguez, 2011] will expose a constant total entropy irreversible process –a contradiction to the law of increasing entropy- existing within the formalism of current second law thermodynamics. This process is nothing more than a by-product of the substitution of the non-zero values assigned by Clausius to the transformations taking place in a reversible cycle, by the zero values of current thermodynamics. In order to locate it we are going to start taking a closer look at the irreversible transfer of heat between two bodies of different temperatures such as that represented in Figure 5.

The essential characteristic of the direct, irreversible transfer of a given amount of heat from a hot to a cold reservoir is that through it no work is at all produced. The fact that the transfer back of this heat from the cold to the hot reservoir demands the expenditure of work, combined with the fact that none was generated in the original transfer, leads to the realization that this reversion can only be accomplished with the concourse of a work-supplying body. This impossibility of recuperating the original universe without changes remaining in other bodies is what qualifies it as irreversible. The fact made evident by figure 5 that such a process involves no other bodies but a hot and a cold reservoir means that they are the only ones changing once the transfer has taken place. Under the assumption that the initial conditions of the hot and cold reservoirs involved in this process are respectively identical to those used in the cyclical process of Figure 1, we can specify the changes experienced by these bodies in the irreversible heat transfer represented in Figure 5, as follows: The final condition of the hot reservoir equals its initial condition minus an amount of heat Q_h; The final condition of the cold reservoir equals its initial condition plus an amount of heat Q_h.

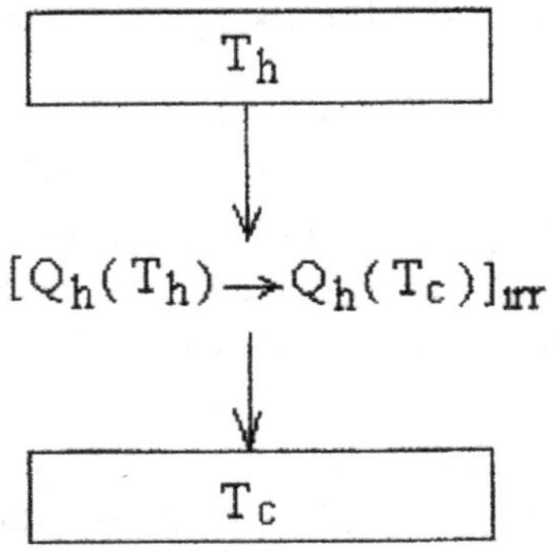

Figure 5. The figure depicts the irreversible transfer of an amount of heat Q_h from a hot reservoir of temperature T_h to a cold reservoir of temperature T_c. The fact that no work is generated in this process makes it irreversible as the transfer back of Q_h from the cold to the hot reservoir can only be achieved by leaving a permanent change on that body supplying the work demanded by such a transfer.

If so, the total entropy change for this process will be given by the combination of the entropy change experienced by the hot reservoir of temperature T_h in releasing an amount of heat Q_h, and the one experienced by the cold reservoir of temperature T_c in absorbing that same amount of heat. Through a direct application of the previously given definition for the entropy function we can determine that those entropy changes amount, respectively, to $-Q_h / T_h$ and Q_h / T_c. Therefore

$$\Delta S[Q_h(T_h) \rightarrow Q_h(T_c)]_{irr} = -(Q_h/T_h) + (Q_h/T_c) \tag{6.1}$$

The re-expression of the previous equation in the following form

$$\Delta S[Q_h(T_h) \rightarrow Q_h(T_c)]_{irr} = Q_h(T_h - T_c)/(T_h T_c) \tag{6.2}$$

makes clear that in this process each and every unit of heat contained in Q_h makes a contribution to the total entropy change in the amount of $(T_h - T_c)/(T_h T_c)$. Let us now recognize that in the previous equation the term $(T_h - T_c)/T_h$ can be identified as the maximum efficiency possible $(\eta_{\max})$ for any heat engine working between those two reservoirs. On this perspective the product $Q_h((T_h - T_c)/T_h)$ represents the maximum possible amount of work (W) this engine could have outputted had Q_h been fed to it. The fact that Q_h was not fed to this engine, but instead it was transferred in a direct, irreversible fashion to the cold reservoir, allows us to realize that in this situation the said product quantifies the wasted work-producing potential i.e. the 'lost work' that Q_h carries with it to the cold reservoir. This last consideration is made explicit in the following re-expression of equation (6.2)

$$\Delta S[Q_h(T_h) \rightarrow Q_h(T_c)]_{irr} = Q_h(T_h - T_c)/(T_h T_c) = Q_h \eta_{\max} / T_c = W / T_c \tag{6.3}$$

6.2.2 An equivalent irreversible heat transfer and the black swan of current thermodynamics

The workless label previously mentioned as the essential characteristic of an irreversible heat transfer opens up the possibility of producing a direct, irreversible transfer of heat from a hot to a cold reservoir via the combination of a first step represented by one cycle in the operation of the reversible cyclical process represented in Figure 1, with a second step represented by the frictional degradation of the work produced by the cycle.

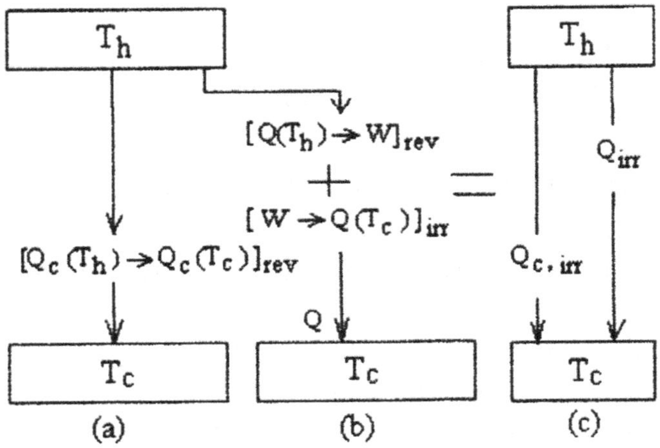

Figure 2'. Out of the amount of heat Q_h released by the hot reservoir of temperature T_h, the reversible engine implied in (a) manages to transfer the smaller amount Q_c to the cold reservoir of temperature T_c, and to transform the difference $Q = Q_h - Q_c$ into the equivalent amount of work W. This amount of work will be now, as shown in (b), frictionally degraded into the equivalent amount of heat Q, and as such it will end up in the cold reservoir. As shown in (c), the concatenation of processes (a) and (b) reduces to the irreversible transformations of the amounts of heat Q and Q_c. The irreversible condition of these two heat transfers is explained by the fact that W is no longer available there. The transfer back of either or both of these amounts of heat from the cold reservoir can only be accomplished by leaving a permanent change in that body called to supply the work demanded by those transfers. In 2'(c) Q_{irr} and $Q_{c,irr}$ stand, respectively, for $[Q(T_h) \rightarrow Q(T_c)]_{irr}$ and $[Q_c(T_h) \rightarrow Q_c(T_c)]_{irr}$

As shown in Figure 2'(a) and as previously discussed, one cycle in the operation of this reversible engine brings forward the transfer of the amount of heat Q_c from the hot to the cold reservoir, as well as the transformation into work of the amount of heat Q of temperature T_h. According to current second law thermodynamics both of these reversible changes or transformations take place, as shown in Table 2, with zero total entropy changes. Let us now take the work generated in 2'(a) and via a frictional process irreversibly degrade it, as shown in 2'(b), into an equivalent amount of heat (Q) that will end up in the cold reservoir. The result of this combination of processes is shown in 2'(c) where we can see that its net effect boils down to the irreversible transfers of the amounts of heat Q and Q_c from the hot to the cold reservoir. The irreversible qualifier stems from the fact that the transfer back of Q or Q_c, or both, from the cold to the hot reservoir demands an amount of work that we don't have. If performed, it will be at the price of leaving a permanent change in that body called to supply the missing work. It should also be noticed that in 2'(c) all the changes brought about by the concatenation of 2'(a) with 2'(b) reside solely in the heat reservoirs. No change remains in the variable body implicit in process 2'(a) as at the end of one cycle we find it in its precise initial condition. No change remains in the work reservoir as the work originally deposited there by process 2'(a) has been retrieved and transformed into heat by process 2'(b). Based on these considerations we can specify the changes experienced by the heat reservoirs on 2'(c) as follows: The final condition of the hot reservoir equals its initial condition minus an amount of heat Q_h; the final condition of the cold reservoir equals its initial condition plus an amount of heat Q_h (It should be remembered here that $Q_h = Q_c + Q$)

The fact that in the processes of Figures 5 and 2'(c) the only bodies affected are the heat reservoirs, combined with the fact that the respective initial and final conditions of the hot and cold reservoirs are the same in both of these processes allows us to conclude that these processes are equivalent, and if so, that the same entropy change applies to both of them. That this is so will be shown via a comparison of the entropy change for the process of Figure 5, as given by equation (6.3), with that obtained for process 2'(c) via the addition of the entropy changes of processes 2'(a) and 2'(b) as shown in equation (6.4)

$$\Delta S[process\,2'(c)] = \Delta S[process\,2'(a)] + \Delta S[process\,2'(b)]$$
$$= \Delta S[Q_c(T_h) \rightarrow Q_c(T_c)]_{rev} + \Delta S[Q(T_h) \rightarrow W]_{rev} + \Delta S[W \rightarrow Q(T_c)]_{irr} =$$

$$0 + 0 + \frac{W}{T_c} = W / T_c \qquad (6.4)$$

All the entropy changes substituted in the previous equation are those sanctioned by current second law thermodynamics, as shown in Table 2. It should also be noted that in Figure 2'(c) the combination of transformations $[Q(T_h) \rightarrow W]_{rev}$ and $[W \rightarrow Q(T_c)]_{irr}$ has been substituted by the expression $[Q(T_h) \rightarrow Q(T_c)]_{irr}$. According to this, the entropy change for this last transformation can be written as follows

$$\Delta S[Q(T_h) \rightarrow Q(T_c)]_{irr} = \Delta S[Q(T_h) \rightarrow W]_{rev} + \Delta S[W \rightarrow Q(T_c)]_{irr} =$$

$$0 + \frac{W}{T_c} = W / T_c \qquad (6.5)$$

The proper combination of equations (6.4) and (6.5) allows us to write the following equation for the entropy change of process 2'(c)

$$\Delta S[process\ 2'(c)] = \Delta S[Q(T_h) \rightarrow Q(T_c)]_{irr} = W / T_c \qquad (6.6)$$

Even if 2'(c) is a process involving the irreversible transfers of both Q and Q_c, from equation (6.6) we learn, however, that its entropy change is *solely and uniquely quantified* by the entropy change associated to the irreversible transfer of Q. This result leads in turn to the inescapable conclusion that the entropy change for the irreversible transfer of the other amount of heat there involved, Q_c, takes place at constant total entropy, i.e.

$$\Delta S[Q_c(T_h) \rightarrow Q_c(T_c)]_{irr} = 0 \qquad (6.7)$$

The realization that the process to which equation (6.7) makes reference is reversible attending to its entropy change, and irreversible attending to the impossibility of transferring Q_c back to the hot reservoir without changes in other bodies remaining, attest to the lack of equivalence between the constant total entropy criterion of reversibility and that embodied by the possibility of restoring the initial condition of the universe. The only possibility open for the constant total entropy reversibility criterion to remain valid in regard to what equation (6.7) expresses, i.e. the only possibility for us to trade '*irr*' for '*rev*' in equation (6.7) demands of Q_c the ability to flow of itself, unassisted, from the cold to the hot reservoir. In other words the validity of the constant total entropy reversibility criterion –and by extension that of the total entropy increase quality of irreversible processes- demands the non-validity of the second law understood as: Heat cannot, of itself, pass from a colder to a hotter body.

The fact that the unassisted transfer of Q_c from the cold to the hot reservoir is denied by experience confirms the irreversibility of $[Q_c(T_h) \rightarrow Q_c(T_c)]_{irr}$, and negates the association between reversibility and a zero entropy change.

On top of all this, the situation depicted in 2'(c) negates the state function condition of the entropy. The fact that a unit of heat in Q_c is no different from a unit of heat in Q, as each of them experience precisely the same change, namely its irreversible transfer from a hot reservoir of temperature T_h to a cold reservoir of temperature T_c, demands the same entropy change for one and the other. The assignment of a zero entropy change for a unit of heat when part of Q_c, and a different from zero entropy change when part of Q, implies that the entropy is not a state function as it assigns different entropy changes to the very same process.

The black swan of current second law thermodynamics takes thus form in transformation $[Q_c(T_h) \rightarrow Q_c(T_c)]_{irr}$ when part of process 2'(c). Against Aristotle's principle of contradiction, the foremost principle of formal logic, this process is reversible and irreversible at the same time and under the same aspect. It is reversible on reason of its entropy change. Irreversible on reason of the fact that its transfer back to the hot reservoir can only be performed at the price of leaving permanent changes in other bodies.

Let us then summarize the conclusions of the previous argument in the following way:

1. That the constant total entropy process represented in (6.7) is irreversible negates the notion subsumed by the law of increasing entropy stating that a zero total entropy change is an exclusive trait of reversible processes.
2. That the constant total entropy process represented in (6.7) is irreversible negates the other notion subsumed by the law of increasing entropy stating that every irreversible process takes place with a positive total entropy change.

The previous results, in disqualifying the law of increasing entropy, call for a revision of the fundamental tenets of what we identify as the second law. This task is to be performed in what follows. The negation of the corollary mentioned at the beginning of this chapter will be one of the results of the argument presented in what follows.

CHAPTER 7: THE CORRECT VALUES FOR THE TRANSFORMATIONS AND THE EMERGENCE OF THE NEGENTROPIC FORMULATION

7.1 THE CORRECT VALUES

The process of Figure 2', the one that allowed for the identification of the logical crack existing in the body of current second law thermodynamics, is the one that will also take us to the correct value for the necessarily reversible transformation of heat into work $[Q(T_h) \rightarrow W]_{rev}$. With this value in our hands we will be in the position of determining the correct value for the reversible transformation of heat from a higher to a lower temperature, $[Q_c(T_h) \rightarrow Q_c(T_c)]_{rev}$.

The combination of the facts that the total entropy change for process 2'(c) is $Q_h(T_h - T_c)/(T_h T_c)$; that $Q_h = Q_c + Q$; and that every unit of heat flowing in 2'(c) from the hot to the cold reservoir is equal to any other, allows us to write the following equations for the entropy changes of the irreversible streams of heat Q_c and Q appearing in 2'(c):

$$\Delta S[Q_c(T_h) \rightarrow Q_c(T_c)]_{irr} = Q_c(T_h - T_c)/(T_h T_c) \qquad (7.1)$$

$$\Delta S[Q(T_h) \rightarrow Q(T_c)]_{irr} = Q(T_h - T_c)/(T_h T_c) \qquad (7.2)$$

As before explained in regard to the transit from eq. (6.4) to eq. (6.5), $[Q(T_h) \rightarrow Q(T_c)]_{irr}$ is shorthand for the concatenation of $[Q(T_h) \rightarrow W]_{rev}$ and $[W \rightarrow Q(T_c)]_{irr}$. Therefore

$$\Delta S[Q(T_h) \rightarrow W]_{rev} + \Delta S[W \rightarrow Q(T_c)]_{irr} = Q(T_h - T_c)/(T_h T_c) =$$

$$-\frac{Q}{T_h}+\frac{Q}{T_c} \qquad (7.3)$$

The fact that $[Q(T_h)\rightarrow W]_{rev}$ is dependent on T_h but not on T_c, while $[W\rightarrow Q(T_c)]_{irr}$ is dependent on T_c but not on T_h, allows us to make the following identifications in equation (7.3)

$$\Delta S[Q(T_h)\rightarrow W]_{rev} = -Q/T_h \qquad (7.4)$$

$$\Delta S[W\rightarrow Q(T_c)]_{irr} = Q/T_c \qquad (7.5)$$

It should be obvious by now that the entropy change for $[W\rightarrow Q(T_h)]_{rev}$, the inverse of the process represented in eq. (7.4), can be written as follows

$$\Delta S[W\rightarrow Q(T_h)]_{rev} = Q/T_h \qquad (7.6)$$

The entropy changes given by equations (7.4), (7.5), and (7.6) constitute the correct values for the transformations there respectively referred. It should be noted in this regard that these three values are identical to those furnished by Clausius' thermodynamics.

Let us note here that the result shown in equation (7.4) negates current thermodynamics' assertion regarding the impossibility of processes taking place with a net decrease in total entropy on reason of any such process being reducible or likened to the spontaneous flow of heat from a colder to a hotter body. In other words, according to current thermodynamics the process of equation (7.4) should be impossible, as impossible as the spontaneous flow of heat from a colder to a hotter body. That this is not so comes from the simple fact that the process behind equation (7.4) is an expansion, a process known to be spontaneous!

Let us also reiterate here the fact that in the process of Figure 2' the degradation of W produces the effect of changing the condition of $[Q_c(T_h)\rightarrow Q_c(T_c)]$ from reversible to irreversible and its entropy change from zero to $Q_c(T_h - T_c)/(T_h T_c)$. As explained before, the transition from reversible to irreversible stems from the fact that once W has been degraded; the transference of Q_c back to the hot reservoir can not be achieved without changes remaining elsewhere. This notion is apparently unknown to current thermodynamics.

7.2 THE REVERSIBLE TRANSFORMATION OF HEAT AND THE 'W' TRANSFORMATION

Armed now with equations (7.4) and (7.6) we will proceed to evaluate the total entropy change of the other transformation produced by a reversible cyclical process: $[Q_c(T_h) \to Q_c(T_c)]_{rev}$.

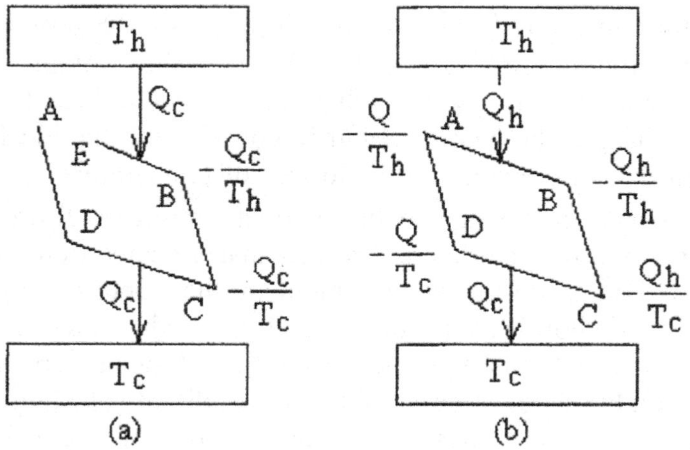

Figure 6. Process (a) represents the concatenation of processes required to produce in a reversible cyclical process, the transformation of heat from one temperature to another. If traveled in the *EBCDA* sequence the heat will be reversibly transformed from the hot to the cold reservoir. If on the sequence *ADCBE,* it will be transferred from the cold to the hot reservoir. Process (b) represents a reversible cyclical process. Both of these processes are used to illustrate the use of the 'W' transformation to calculate the correct entropy changes for the two transformations there taking place. The fractions appearing at selected points of these graphs form part of these calculations.

Even if this transformation is the result of concatenation *EB*, *BC*, *CD*, and *DA* as represented in Figure 6(a), our attention will be, as has been done for all of our previous analysis regarding this transformation, solely focused on isothermal and reversible processes *EB* and *CD*; Again, the reason for this being that the effects of process *BC* are precisely cancelled by the effects of process *DA*.

Through isothermal and reversible expansion *EB* the ideal gas absorbs from the hot reservoir an amount of heat Q_c and transforms it into an equivalent amount of work W_c. In doing so process *EB* brings forward transformation $[Q_c(T_h) \rightarrow W_c]_{rev}$ with an associated entropy change, as equation (7.4) indicates, of $-Q_c / T_h$. If we assume a value of zero for the entropy of the universe at point *E*, then the entropy at point *B* will be $S_B = -Q_c / T_h$. Let us now agree that process *BC* not only changes the volume and temperature of the gas from their values at *B* to those at *C*; it also changes the heat reservoir in contact with the ideal gas from that of temperature T_h to that of temperature T_c. The mechanical reservoir and the work it safe-keeps form now part of the new situation. The fact that the conditions of all these bodies define at any moment the condition of the universe allows us to understand that what process *BC* actually accomplishes is the substitution of the universe from the conditions existing at *B*, to the conditions existing at *C* and in doing so brings forward an entropic effect consisting in shifting the ΔS associated to the production of W_c out of heat of temperature T_h, for that which, in the amount of $-Q_c / T_c$, would have arisen had the same amount of work been generated from heat at the new temperature T_c. The rationale is actually very simple. Once W_c and the rest of the universe arrive at C, this work is indistinguishable from a similar amount generated at T_c. The entropy change produced by the substitution of universe *B* by universe *C* amounts to $\Delta S[B \rightarrow C] = (-Q_c / T_c) - (-Q_c / T_h)$. This effect, unrecognized up to now on reason of the zero values assigned by current thermodynamics to the reversible inter-conversions between heat and work, will be here be called the 'W' transformation, and represented either as $[W(T_1) \rightarrow W(T_2)]$ or as *[WT].* In general, the entropy change of the 'W' transformation can be written as follows

$$\Delta S[W(T_1) \rightarrow W(T_2)]_{rev} = (-Q / T_2) - (-Q / T_1) \qquad (7.7)$$

The '*W*' Transformation operates on the entropy (actually, the negentropy) associated to work produced along isothermal and reversible processes, or to work remaining once an isothermal an reversible compression has taken place. No work remaining means no negentropy for the '*W*' Transformation to operate on and consequently a zero entropy change for this effect. Continuing with the concatenation being considered, let us agree that the universe finds itself now at *C*. The entropy at this point can be calculated as the summation of the entropy at point *B* plus the entropy change produced by the '*W*' Transformation along *BC* i.e. $S_C = S_B + \Delta S[W_c(T_h) \rightarrow W_c(T_c)]$. Substitution of $-Q_c / T_h$ for S_B, and $(-Q_c / T_c) - (-Q_c / T_h)$ for $\Delta S[W_c(T_h) \rightarrow W_c(T_c)]$ leads to $S_C = -Q_c / T_c$. The next process of the concatenation is represented by the isothermal and reversible compression *CD*. Here an amount of work W_c identical to that available in the mechanical reservoir is consumed along this process, and the spent work released as heat to the cold reservoir. In doing so process *CD* brings forward transformation $[W_c \rightarrow Q_c(T_c)]_{rev}$ with an associated entropy change of Q_c / T_c. Notice that at the conclusion of this process no more work is available in the mechanical reservoir. The entropy of the universe at point *D* can now be evaluated as the summation of its entropy at *C*, plus the entropy change along *CD*. When this is done we get: $S_D = (-Q_c / T_c) + (Q_c / T_c) = 0$. The entropy of the universe at point D happens to be zero and thus identical to that assumed for the initial point of the concatenation, point E. This result is logical and explainable for the following reasons: First, at D no work remains. No work, no negentropy. Second, that all the other processes taking place from E to D are heat transfers between bodies of essentially the same temperature, all of which take place at constant total entropy. Let us now consider process DA, the last leg of the concatenation defining the reversible transformation of Q_c from the hot to the cold reservoir, and let us agree that since no work and consequently no entropy (actually, negentropy) remains at D, the 'W' Transformation has nothing to operate on and on reason of this such a transformation produces no effect along DA, i.e. its contribution along this path is zero. Therefore, $S_A = S_D + \Delta S[DA] = 0 + 0 = 0$. We are now in the position to calculate the total entropy change for concatenation $EBCDA$, or in other words, for the reversible transformation of Q_c from the hot to the cold reservoir. This calculation, taking the form [entropy change = final entropy - initial entropy] leads to $\Delta S[EBCDA]_{rev} = S_A - S_E = 0 - 0 = 0$. If so, then

$$\Delta S[Q_c(T_h) \rightarrow Q_c(T_c)]_{rev} = 0 \qquad (7.8)$$

The previous description will be formally presented, with reference to Figure 6(a), as follows:

$$\Delta S[Q_c(T_h) \rightarrow Q_c(T_c)]_{rev} = \Delta S[EB] + \Delta S[WT]_{BC} + \Delta S[CD] + \Delta S[WT]_{DA}$$

(7.9)

$$\Delta S[EB] = \Delta S[Q_c(T_h) \rightarrow W_c]_{rev} = -Q_c / T_h \qquad (7.10)$$

$$\Delta S[WT]_{BC} = \Delta S[W_c(T_h) \rightarrow W_c(T_c)]_{rev} = (-Q_c / T_c) - (-Q_c / T_h) \qquad (7.11)$$

$$\Delta S[CD] = \Delta S[W_c \rightarrow Q_c(T_c)]_{rev} = Q_c / T_c \qquad (7.12)$$

$$\Delta S[WT]_{DA} = 0 \qquad (7.13)$$

Therefore

$$\Delta S[Q_c(T_h) \rightarrow Q_c(T_c)]_{rev} = -\frac{Q_c}{T_h} - \frac{Q_c}{T_c} + \frac{Q_c}{T_h} + \frac{Q_c}{T_c} = 0 \qquad (7.14)$$

This is:

$$\Delta S[Q_c(T_h) \rightarrow Q_c(T_c)]_{rev} = 0 \qquad (7.15)$$

And, necessarily

$$\Delta S[Q_c(T_c) \rightarrow Q_c(T_h)]_{rev} = 0 \qquad (7.16)$$

At this point the fundamental set of values of the corrected version is complete. This new construction is to be heretofore called the Negentropic Formulation (*NF*) on reason of the fact that it has allowed us to realize that negentropic thermodynamic processes, this is, processes taking place with a negative total entropy change –like the transformation of heat into work- are indeed possible. The main results associated to this formulation will be eventually subjected, along those of Clausius and *CT*, to the surgical knife of dialectics, as embodied in its three fundamental laws.

Serves to point out here that identical values to those given by equations (7.4) and (7.8) were previously obtained by the author through a different argument involving the connection between lost work and entropy production, as well the notion that work is a higher quality form of energy than heat [Íñiguez, 2008]

In order for the reader to be able to compare the values that these three different formulations of the second law assign to the essential transformations involved in thermodynamic processes, they are presented tête-à-tête in Table 3.

With the corrected values in place we can address the problem of determining the total (universe) entropy change for a reversible cyclical process. This will be done via the summation of the entropy changes for the two transformations produced by it, i.e.

$$\Delta S_{NF}[Cycle]_{rev} = \Delta S[Q(T_h) \rightarrow W]_{rev} + \Delta S[Q_c(T_h) \rightarrow Q_c(T_c)]_{rev} \quad (7.17)$$

When the correct values for these transformations, $-Q/T_h$ and zero, respectively, are substituted in eq. (7.17), the following result is obtained

$$\Delta S_{NF}[Cycle]_{rev} = -\frac{Q}{T_h} + 0 = -\frac{Q}{T_h} \quad (7.18)$$

Equation (7.18) shows that contrary to current thermodynamics, Carnot's reversible engine is actually a negentropic process, with the entropy decrease proportional to its work output.

Transformation	Clausius	CT	NF
$[Q(T_h) \rightarrow W]_{rev}$	$-\frac{Q}{T_h}$	0	$-\frac{Q}{T_h}$
$[W \rightarrow Q(T_h)]_{rev}$	$\frac{Q}{T_h}$	0	$\frac{Q}{T_h}$
$[W \rightarrow Q(T_h)]_{irr}$	$\frac{Q}{T_h}$	$\frac{Q}{T_h}$	$\frac{Q}{T_h}$
$[Q_c(T_h) \rightarrow Q_c(T_c)]_{rev}$	$Q_c(\frac{1}{T_c} - \frac{1}{T_h})$	0	0
$[Q_c(T_c) \rightarrow Q_c(T_h)]_{rev}$	$Q_c(\frac{1}{T_h} - \frac{1}{T_c})$	0	0
$[Q_c(T_h) \rightarrow Q_c(T_c)]_{irr}$	$Q_c(\frac{1}{T_c} - \frac{1}{T_h})$	$Q_c(\frac{1}{T_c} - \frac{1}{T_h})$	$Q_c(\frac{1}{T_c} - \frac{1}{T_h})$

Table 3. The set of values for the Negentropic Formulation (NF) is here shown vis-à-vis those of Clausius and Current Thermodynamics (CT). Note that in the Negentropic Formulation the only negentropic process is the transformation of heat into work.

The result shown by eq. (7.18) is a manifestation of the fact that at the end of one cycle in the operation of this engine, a non-compensated change has been left in the universe in the form of an amount of work that previously wasn't there. The uncompensated qualifier comes as a result of the fact that in a reversible cycle the entropy changes for the reservoirs are indeed compensated –they add up to zero-. It is on this perspective that work production becomes an uncompensated change in those processes.

Given the radically new perspective brought about by this new notion represented by the 'W' transformation, an analysis similar to that leading to equation (7.18) will be carried on the cycle of Figure 6(b) with the purpose of determining its total entropy change.

Let us then write the total or universe entropy change in terms of processes $AB, [WT]_{BC}$, CD, and $[WT]_{DA}$. Then

$$\Delta S_{NF}[Cycle]_{rev} = \Delta S[AB] + \Delta S[WT]_{BC} + \Delta S[CD] + \Delta S[WT]_{DA} \tag{7.19}$$

$$\Delta S[AB] = \Delta S[Q_h(T_h) \to W_h]_{rev} = -Q_h / T_h \tag{7.20}$$

$$\Delta S[WT]_{BC} = \Delta S[W_h(T_h) \to W_h(T_c)]_{rev} = -(Q_h / T_c) - (-Q_h / T_h) \tag{7.21}$$

$$\Delta S[CD] = \Delta S[W_c \to Q_c(T_c)]_{rev} = Q_c / T_c \tag{7.22}$$

Let us now note that of the amount of work W_h available to carry on compression CD, only a portion of it in the amount W_c is actually used, leaving a work remnant at D in the amount of $W = W_h - W_c$. The 'W' transformation operating along DA is to change or transform the negentropy carried on by W as part of the cold isotherm, into that it conveys as part of the hot isotherm, i.e.

$$\Delta S[WT]_{DA} = \Delta S[W(T_c) \to W(T_h)]_{rev} = (-Q / T_h) - (-Q / T_c) \tag{7.23}$$

The addition of equations (7.20)-(7.23) produces

$$\Delta S_{NF}[Cycle]_{rev} = -\frac{Q_h}{T_h} - \frac{Q_h}{T_c} + \frac{Q_h}{T_h} + \frac{Q_c}{T_c} - \frac{Q}{T_h} + \frac{Q}{T_c} = -\frac{Q}{T_h} \tag{7.24}$$

$$\Delta S_{NF}[Cycle]_{rev} = -\frac{Q}{T_h} \qquad (7.25)$$

The equality of the results shown by equations (7.18) and (7.25) is a testament of the self-consistency of the negentropic formulation.

The '*NF*' sub-index used previously to identify this new formulation will be used no more. From now on the entropy change for the universe of a given process evaluated via the negentropic formulation, will be simply symbolized as ΔS_u, the sub index '*u*' indicating "universe'.

Note that the addition of zero, in the form of $-\frac{Q_h}{T_h}+\frac{Q_c}{T_c}$, to equations (7.18) and (7.25) allows writing the following expression for the reversible cycle

$$\Delta S_u[Cycle]_{rev} = -\frac{Q_h}{T_h}+\frac{Q_c}{T_c}-\frac{Q}{T_h} \qquad (7.26)$$

7.3 A SUMMARY OF THE TWO PREVIOUS SECTIONS

In subsection 6.2.2 it was shown that current thermodynamics allows in its realm the existence of a constant total entropy process that is both, reversible on reason of its zero entropy change, and irreversible on reason of the impossibility of recuperating the original condition of the universe without changes remaining elsewhere. The mere existence of this impasse, i.e. the mere fact of current thermodynamics allowing the existence of a process that is both, reversible and irreversible at the same time and under the same aspect, constitutes irrefutable evidence of its fallacious nature. The solution of the impasse allowed us to calculate a total entropy change of $-Q/T$ for the necessarily reversible transformation into work of an amount of heat Q

of temperature T; and in doing so brought forward the realization that negentropic processes –processes occurring with a total negative entropy change- were indeed possible. A simple logical extension of this result led us to a value of Q/T for the opposite process, this is, for the reversible transformation of work into an equivalent amount of heat of temperature T. These two values led us in turn, through the 'W' Transformation, to a value of zero for the reversible transformation of heat taking place in a reversible cyclical process, and this, in turn to a total negative entropy change for the reversible cycle. These results made clear the fact that reversible processes do not necessarily take place at constant total entropy. The total entropy changes for the reversible cycle, as well as transformations $[Q(T_h) \rightarrow W]_{rev}$, and $[W \rightarrow Q(T_h)]_{rev}$ attest to this, and in doing so negate the validity of the constant total entropy reversibility criterion. The true criterion of reversibility –if there ever was another one- is Planck's criterion centered on the recuperation of the original universe. This criterion is true beyond doubt simply because it is actually a re-expression of that "statement of experimental truth" [Schmidt, 1966, p. 92] that says that heat cannot, of itself, pass from a colder to a hotter body. The fact that any irreversibility can be reduced to an irreversible flow of heat from the hot to the cold reservoir means that the only path open for the recuperation of the initial condition of the universe, without changes remaining elsewhere, demands of this heat the ability to flow of itself from the colder to the hotter body. Since this is impossible, it is also impossible to recover the initial condition of the universe once an irreversible process has taken place. But Planck's criterion imposes no requirement, no obligation, no restriction nor condition demanding a particular value for the total entropy change of a reversible process, a position parallel to that of the negentropic formulation. Nowhere does Planck statement imposes on the total entropy change of $[A \rightarrow B]$ the obligation to assume this or that value if evolving reversibly.

Here the bracket, as in any other transformation, represents all the bodies involved in the process under scrutiny. What it clearly states is that if the original condition of the universe can be restored once the process has taken place, without changes in other bodies remaining, then the said process is reversible; if not, irreversible. Since entropy is a state function, and the universe is being restored to its precise original condition, without any other additional change, then

$$\Delta S\{[A \to B]+[B \to A]\}_{rev} = 0 \qquad (7.27)$$

And from it we learn that the only subsidiary condition to the return of the universe to its initial condition is represented by

$$\Delta S[A \to B] = -\Delta S[B \to A] \qquad (7.28)$$

As should be obvious, the previous condition can indeed be satisfied by any real number whatsoever.

If reversibility is not tied to a particular value of the total entropy change, it might even be possible for the reversible and irreversible versions of a given process to carry the same values for the total entropy change. That one is reversible and the other irreversible cannot be ascertained from the values themselves, but from the possibility of returning their respective universes to their original conditions, without changes remaining elsewhere.

CHAPTER 8: IRREVERSIBLE CYCLICAL PROCESSES

8.1 A FIRST APPROACH

The argument presented in this section was part of previous publications by the author [Íñiguez, 1999 and 2008]

The problem of finding the negentropic formulation's expression for the entropy change of the universe of an irreversible cyclical process will be now addressed. Let us consider the cyclical process depicted in Figure 7(b).

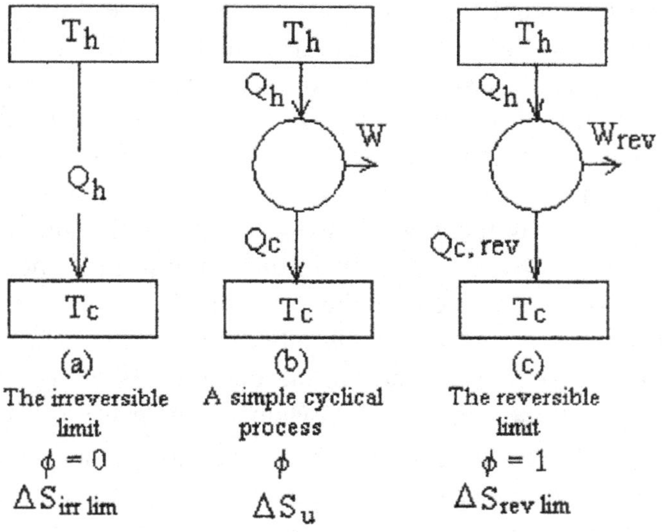

Figure 7. A simple cyclical process (b), and its limiting reversible (c), and irreversible (a) operations

In it Q_h, Q_c and W stand, respectively, for the amount of heat given off by the hot reservoir at T_h, taken in by the cold reservoir at T_c, and transformed into work. The limits to the operation of this process are represented in Figures 7(a) and

7(c). In the reversible limit shown in (c) the cyclical process produces the maximum amount of work possible (W_{rev}) from the amount of heat absorbed from the hot reservoir (Q_h), discarding consequently to the cold reservoir the minimum possible amount of heat ($Q_{c,rev}$). In the irreversible limit on the other hand, all of the heat given off by the hot reservoir finds its irreversible way to the cold reservoir; consequently, no work is here at all produced. In the reversible limit the simple performance of the inverse cycle can restore the universe of this process to its initial condition, without changes remaining elsewhere. In the irreversible limit, on the other hand, no part of the universe is susceptible of being restored to its initial condition without changes in other bodies remaining. Any of the cyclical processes bound by these limit operations are thus *partially reversible* in the sense that in any of them only a portion of their universe is susceptible of being restored to its initial condition, this portion being that involved in the production of whatever amount of work the said process was able to output in the first place. This can be understood if it is realized that by feeding this work to a reversible refrigerator, it is possible to restore to its initial condition that portion of the universe involved in the production of the said work, without the occurrence of any other permanent change. Since the whole of the universe is susceptible of being restored when $W = W_{rev}$, while non is when $W = 0$, it follows that the above referred portion i.e. the reversible fraction of the original universe will be given by W / W_{rev}. This quotient will be subsequently referred to as the *reversibility degree,* and symbolized as ϕ. Obviously, the limit values for ϕ are, respectively, zero and one at the irreversible and reversible limits previously described.

It should not be difficult for the reader to realize that an equivalent definition to the one just given can be written for ϕ as the quotient of the efficiency of the cyclical process and that of its reversible limit, as follows: $\phi = \eta / \eta_{rev}$. This alternate definition sheds light on the fact that ϕ is ultimately a 'normalized' or 'reduced' kind of efficiency.

The entropy change for our cyclical process (b) will thus be bound on one side by the entropy change of the irreversible limit (*irr lim*):

$$\Delta S_u[\phi = 0] = \Delta S_{irr\,\lim} = Q_h(T_h - T_c)/(T_h T_c) \tag{8.1}$$

And on the other by the entropy change of the reversible limit (*rev lim*):

$$\Delta S_u[\phi = 1] = \Delta S_{rev\,\lim} = -\frac{W_{rev}}{T_h} \tag{8.2}$$

The contributions these limiting operations make to the total entropy change (ΔS_u) of the cyclical process depend on the relative weight each of those extremes has in the actual process. The reversible contribution to ΔS_u, determined by the amount of work W produced by this cyclical process, will thus be given by $-W/T_h$, which can also be written as $(W/W_{rev})(-W_{rev}/T_h)$, or as $\phi\,\Delta S_{rev\,\lim}$. The heat being irreversibly transferred from the hot to the cold reservoir will be quantified as follows: That an amount of work W is produced, with $W \langle W_{rev}$, means that of all the heat given off by the hot reservoir only the portion $(Q_h/W_{rev})W$ is used for the production of W. The rest: $Q_h - (Q_h/W_{rev})W$ or $(1-\phi)Q_h$ is consequently that being irreversibly transferred; and to this amount corresponds an entropy change of $(1-\phi)Q_h(T_h - T_c)/T_h T_c$, which can also be written as $(1-\phi)\Delta S_{irr\,\lim}$. At the light of the results just given the ΔS_u expression for our cyclical process can be written as follows:

$$\Delta S_u = (1-\phi)\Delta S_{irr\,\lim} + \phi\,\Delta S_{rev\,\lim} \tag{8.3}$$

Or as:

$$\Delta S_u = (1-\frac{W}{W_{rev}})Q_h(\frac{T_h - T_c}{T_h T_c}) - \frac{W}{T_h} \tag{8.4}$$

The proper algebraic manipulation of the first term of the right hand side of equation (8.4), which includes the following steps: $\frac{W_{rev} - W}{Q_h \eta_{rev}} \cdot \frac{Q_h \eta_{rev}}{T_c}$, $\frac{Q_c}{T_c} - \frac{Q_{c,rev}}{T_c}$, followed by the addition and subtraction of Q_h / T_h as well as making use of the fact that $(-Q_h / T_h) + (Q_{c,rev} / T_c) = 0$, leads to:

$$(1-\phi)\Delta S_{irr\,\lim} = (1-\frac{W}{W_{rev}})Q_h(\frac{T_h - T_c}{T_h T_c}) = -\frac{Q_h}{T_h} + \frac{Q_c}{T_c} \tag{8.5}$$

The substitution of W by Q on the other term of equation (8.4) leads, on its part, to

$$\phi\, \Delta S_{rev\,\lim} = -\frac{Q}{T_h} \tag{8.6}$$

The substitution of equations (8.5) and (8.6) in (8.4) leads to

$$\Delta S_u = -\frac{Q_h}{T_h} + \frac{Q_c}{T_c} - \frac{Q}{T_h} \tag{8.7}$$

The agreement between equations (7.26) and (8.7) is yet another indication of the self-consistency of the negentropic formulation. Equation (8.7) represents this way the general expression furnished by the negentropic formulation for the total entropy change of cyclical processes, reversible or not.

It should be noted that equation (8.7) reduces to its two parent equations when the proper conditions are in it inserted. If the cyclical process is reversible ($\phi = 1$) then the summation of the first two terms on the right hand side becomes zero; Q becomes W_{rev}; and equation (8.7) becomes equation (8.2). If the process is irreversible ($\phi = 0$), Q

becomes zero; $Q_c = Q_h$; and equation (8.7) becomes equation (8.1).

Through its antecedents, equations (8.5) and (8.6), it becomes clear that equation (8.7) subsumes the fact that the entropy change associated to the universe of a given cyclical process is constituted by a reversible, and an irreversible contribution. Introduction here of the following variable changes

$$\Delta S_{irr} = (1-\phi)\Delta S_{irr\,\lim} \qquad (8.8)$$

$$\Delta S_{rev} = \phi\,\Delta S_{rev\,\lim} \qquad (8.9)$$

Allows us to re-express equation (8.3) as follows

$$\Delta S_u = \Delta S_{irr} + \Delta S_{rev} \qquad (8.10)$$

The fact

$$\Delta S_{irr} = -\frac{Q_h}{T_h} + \frac{Q_c}{T_c} \qquad (8.11)$$

permits the immediate identification of ΔS_{irr} with what both Clausius and current thermodynamics identify as the universe entropy change. Here, however, this term has become the irreversible component of the universe entropy change; as such this term adopts non-zero values only in the case of irreversible processes. Its effect disappears at the reversible limit where its contribution is effectively zero.

The fact that

$$\Delta S_{rev} = -\frac{W}{T_h} \qquad (8.12)$$

Permits identifying this term, the reversible component of the universe entropy change, as that separating the negentropic formulation from the old formulations. As long as work is produced from heat there will be a negative contribution to the universe entropy change. Notice that in all those cyclical processes in which no work is at all produced we have that

$$\Delta S_{rev} = 0 \tag{8.13}$$

In this situation

$$\Delta S_u = \Delta S_{irr} \tag{8.14}$$

And the negentropic formulation collapses into the old formulations. This result allows us to realize that it is at the irreversible limit ($\phi = 0$) where current thermodynamics reigns supreme. The fact shown in Table 3 that the only negentropic transformation is that associated to the conversion of heat into work, combined with the fact that such a transformation plays no role at the irreversible limit, will allow us to understand that every process at this end of the efficiency spectrum is entropic because each and every transformation by these processes subsumed is also entropic.

It is important to realize that as shown in eq. (8.7), the negentropic effect of the heat to work transformation (or the entropic effect of the transformation of work into heat in a refrigerator) appears or manifests itself as a separate, distinct entropic effect from those of the reservoirs. At difference of Clausius formulation which considered the entropic effect of the heat to work transformation as subsumed by that of the hot reservoir, here we can see that it is actually independent of it.

As it was previously discussed in reference to Figure 7, the entropy change for the universe of a cyclical process transits from a positive to a negative value as we move from the irreversible to the reversible limit. Being this so, it should not be difficult to realize that there must exist an operation intermediate between the two extremes just described for which $\Delta S_u = 0$. The fact that it is the efficiency of the operation $\eta = (W/Q_h) = (Q/Q_h)$ the magnitude that allows us to discriminate any one of these operations from any other, combined with the fact that for the operative limits just described the efficiency is respectively zero and $\eta_{rev} = W_{rev}/Q_h = (T_h - T_c)/T_h$, allows us to rephrase our previous statement by saying that there exists an operation of efficiency η_{umbral}, where $0 < \eta_{umbral} < \eta_{rev}$, for which $\Delta S_u = 0$. The name originates in the fact that operations with efficiencies smaller than this one are entropic, while those with larger efficiencies are negentropic. Substitution of condition $\Delta S_u = 0$ in eq. (8.7) followed by adequately solving for Q/Q_h, leads to the identification of the following expression for η_{umbral} :

$$\eta_{umbral} = \frac{T_h - T_c}{T_h + T_c} \qquad (8.15)$$

The relation between the umbral and reversible efficiencies starts by writing the following inequality $(T_h - T_c)/(T_h + T_c) < (T_h - T_c)/T_h$ which assumes that the former is smaller than the latter. Dividing both sides of the inequality by $(T_h - T_c)$ followed by inversion of the respective resulting fractions leads to $T_h + T_c > T_h$. The fact that this last statement is true for all $T_c > 0$ proves the correctness of the original assumption, and on reason of this we state that $\eta_{umbral} < \eta_{rev}$.

8.2 A SECOND APPROACH

Let us consider the situation depicted in Figure 8. In it the simple irreversible cyclical process represented in (b) has been separated into its reversible component, represented in (d); and its irreversible component represented in (e). Out of the total amount of heat Q_h coming out of the hot reservoir, this simple cyclical process manages to produce an amount of work W smaller than $W_{rev} = Q_h(T_h - T_c)/T_h$. In (d) Q_h^* and Q_c^* represent, respectively, the amount of heat coming out of the hot reservoir, and taken in by the cold reservoir, had the production of W taken place in a reversible fashion.

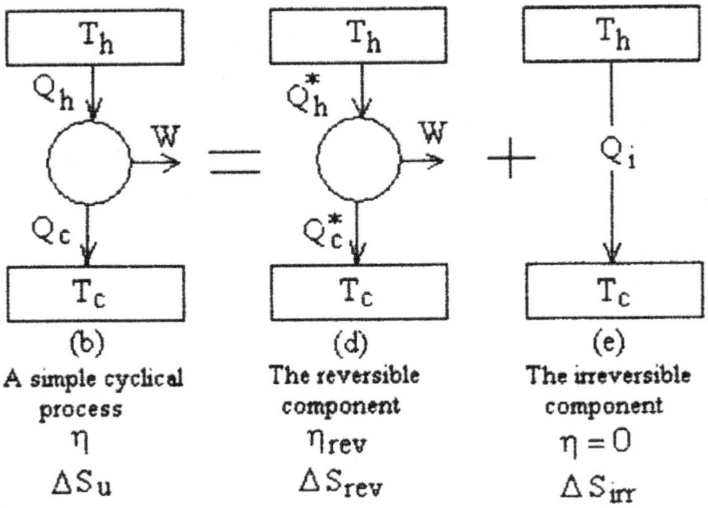

Figure 8. A simple cyclical process of efficiency η is here separated into its reversible and irreversible components. The entropy change for this process is determined by the summation of the contributions of its components

The irreversible component depicted in (e) shows the amount of heat Q_i flowing in a direct, irreversible fashion from the hot to the cold reservoir, bypassing this way the work-producing circuit of the engine. In doing this Q_i carries with it to the cold reservoir an untapped work producing potential, the lost work, in the amount of $W_{lost} = Q_i(T_h - T_c)/T_h$. This amount of heat irreversibly flowing from the hot to the cold reservoir can be evaluated as $Q_i = Q_h - Q_h^*$, or as

$Q_i = Q_c - Q_c^*$. It should be clear to the reader that reversible will be any cycle for which $Q_i = 0$.

The separation shown in Figure 8 brings forward a different but equivalent route to that presented in the previous section in order to determine the entropy change for the universe of irreversible cyclical process. In terms of the entropy changes of its reversible and irreversible components, the entropy change for the cyclical process shown in 8(b) can be written as follows

$$\Delta S_u = \Delta S_{irr} + \Delta S_{rev} \tag{8.16}$$

According to the notation of Figure 8 this equation can also be expressed as follows

$$\Delta S_u = Q_i(\frac{T_h - T_c}{T_h T_c}) - \frac{W}{T_h} \tag{8.17}$$

For process 8(b) we can write: $W/\eta = Q_h$, and $W_{rev}/\eta_{rev} = Q_h$, while for process 8(d): $W/\eta_{rev} = Q_h^*$. The proof of the equivalence of equations (8.17) and (8.4), as a simple inspection of these two equations makes evident, passes through the proof that $Q_i = (1 - \frac{W}{W_{rev}})Q_h$. That this is so starts by putting Q_i equal to $Q_h - Q_h^*$, followed by a simple factorization. In doing this we get $Q_i = Q_h(1 - \frac{Q_h^*}{Q_h})$. If in this expression Q_h and Q_h^* are substituted by their defining equations given above, we get the following expression $Q_i = Q_h(1 - \frac{W/\eta_{rev}}{W_{rev}/\eta_{rev}})$ which, on further simplification, reduces to $Q_i = (1 - \frac{W}{W_{rev}})Q_h$, QED.

It should here be obvious that if (8.17) is equivalent to (8.4), it is also equivalent to (8.7). The same entropy change is produced with independence of whether the effects of a cyclical process are expressed in terms of its limit operations, or in terms of its reversible and irreversible components. For current thermodynamics, the entropy change for process 8(b) is given by

$$\Delta S_{CT} = -\frac{Q_h}{T_h} + \frac{Q_c}{T_c} = -\frac{(Q_h^* + Q_i)}{T_h} + \frac{(Q_c^* + Q_i)}{T_c} = (-\frac{Q_h^*}{T_h} + \frac{Q_c^*}{T_c}) + \frac{Q_i(T_h - T_c)}{T_h T_c} =$$

$$0 + \frac{W_{lost}}{T_c} = \frac{W_{lost}}{T_c} \qquad (8.18)$$

The correct entropy change, however, that of the negentropic formulation, is the following

$$\Delta S_u = (-\frac{Q_h}{T_h} + \frac{Q_c}{T_c}) - \frac{W}{T_h} = \frac{W_{lost}}{T_c} - \frac{W_{gained}}{T_h} \qquad (8.19)$$

In the previous equation work gained means work produced. It should also be noted in equation (8.19) that on reason of being $T_h \rangle T_c$ the entropic effect produced by every unit of work lost is larger than the negentropic effect associated to every unit of work gained. If the image is allowed it may very well be said that the punishment for losing work is larger than the reward for gaining it.

Keeping in mind the admonishment of McGlashan [1966], we will skip the temptation of using equation (8.19) to pontificate about the fate of the universe as a whole, directing our attention instead to its connection with the terrestrial affairs centered on human organizations and their evolution. In this context the message of eq. (8.19) can be interpreted as the entropy of human societies being the resultant of the interplay between chaos and order, between progress and regress. This perception finds sustenance in the fact that what appears as work gained in the previous equation comes as the product of increased efficiencies in the production as well as the utilization of the ordered energy we call work, and because of this it also represents increased access, or better, access to increased levels of negentropy, order, structure, organization and development. This is the work which through the creative powers of man's brain and its coupling with the transforming powers of man's hands has given us knowledge, techniques, and through them the possibility of improving the quality of life for increased numbers of human beings. The proper re-investment of gained work can be found behind the appearance of the wheel, the lever, the pulley, the development of agriculture and metal working, re-investment which in a truly dialectical fashion allowed us the access to increasing amount of gained work. It has been this increased access to gained work coupled with an increased efficiency in its utilization what has taken human societies through that path recognized as progress, development. This is what has led us to increased amounts and quality of foodstuffs, to schools, research centers, hospitals, vaccines, books, motors, cars, airplanes, boats, tractors, fertilizers, spaceships, the telephone, the computer, etc. In summary, it is gained work what propels human societies along the course of progress.

The lost work appearing in eq. (8.19) includes of course the tax nature imposes on us for the production of gained work, including that of maintaining our energy transforming engines in working condition, i.e. those losses and reinvestments of work necessarily accompanying our energy generation processes at each stage of technological advance. It is however the work lost beyond the one previously described the one that that in the human perspective can be associated with regress, with chaos and disorganization. It can be thought as that chunk of W_{gained} that in a manner similar to that represented in Figure 2(b), is generated only to be squandered. This the work that ends up propelling the machinery of consumerism, the same that provides all the gadgetry with which the bored denizen of so-called industrialized –and not so- nations escape the futility of a life whose success is measured not by the constructive contributions made to society, however modest these contributions might be, but by how much stuff can be bought, or how many material things can be possessed. This is also the work used to propel the military machinery, the same that manufactures guns, assault rifles, missiles, nuclear weapons as well as that lost in initiating and maintaining wars; but beyond this, it also represents the waste of human lives and human minds through famine, underdevelopment, corruption, illiteracy, disease, discrimination, pollution, drugs, usury; It also takes form in the dilapidation of human potential brought upon by the rampant stupidity of those governments curtailing, inhibiting the development of the creative powers of their population, as well using religion, gender, drugs, race, language, 'music', 'literature' and similar means to ignite and/or enhance barbaric, animalistic patterns of behavior on their populations. It is the relative weight of this lost work in comparison with the previously described gained work what, in the philosophical perspective of equation (8.19), separates entropic from negentropic societies. It serves to mention that those societies incapable of recognizing these facts face, inevitably, through the continuous misuse of its gained work, an inevitable regress from order to chaos, a situation currently being faced by more than one society on this Earth.

A shorter and definitely more eloquent rendition of the above ideas has been given by Eduardo Césarman, as follows:

La entropía, como una fuerza prevalente, se hace sentir cuando la violencia triunfa sobre la paz, el odio sobre el amor, la locura sobre la razón, la enfermedad sobre la salud, la miseria sobre la abundancia, la muerte sobre la vida, la ignorancia sobre el conocimiento, la necedad sobre la sabiduría y la mentira sobre la verdad.
[E. Césarman, 1997, Hombre y Entropía, p. 17]

8.3 A NOTE ON KELVIN'S *DICTUM* ABOUT THE IMPOSSIBILITY OF RESTORATION OF MECHANICAL ENERGY

La magia y el mito surgen cuando la razón claudica [E. Césarman, Orden y Caos, 1997, p. 15]

William Thomson (Lord Kelvin) was the first to point out that the second law of thermodynamics subsumed a somber statement about the fate of the universe, and of man with it. In his own words

1. There is at present in the material world a universal tendency to the dissipation of mechanical energy.

2. Any restoration of mechanical energy, without more than an equivalent of dissipation, is impossible in inanimate material processes, and is probably never effected by means of organized matter, either endowed with vegetable life or subject to the will of an animated creature.

3. Within a finite period of time past, the earth must have been, and within a finite period of time to come the earth must again be, unfit for the habitation of man as at present constituted, unless operations have been, or are to be performed, which are impossible under the laws to which the known operations going on in the material world are subject.

W. Thomson, 1852

The analysis and refutation of Thomson's *dictum* will be carried on in what follows. Let us refer again to Figure 8. As already explained, process 8(b) is a cyclical process of efficiency η. Its reversible and irreversible components are those shown in 8(d) and 8(e), respectively. The fact that any reversible process produces the maximum possible amount of work at its particular conditions of operation, allows us to conclude that no work is lost in any such process. The previous consideration can be re-expressed by saying that the heat discarded to the cold reservoir by a reversible cyclical process carries with it no unused or wasted work-producing potential. Any claim in the contrary will imply that efficiencies larger than the reversible efficiency are possible, or that reversible process can output work in excess of W_{rev}. The situation is different, however, for process 8(e). Its total entropy change in the amount of

$$\Delta S[8(e)]_{irr} = Q_i(T_h - T_c)/(T_h T_c) = (Q_i(T_h - T_c)/T_h)/T_c = \frac{W_{lost}}{T_c} \qquad (8.20)$$

is an indication of the fact that in its irreversible transit Q_i carries with it to the cold reservoir an amount of wasted, unused work producing potential equal to W_{lost}.

Let us now assume that once a cycle in the operation of process 8(b) has taken place, we are to attempt to restore to the hot reservoir all of the heat Q_c discarded to the cold reservoir. Look again at process 8(d) and ask yourself what would happen if the work W there generated is now fed to the precise inverse of this reversible engine, i.e. to a reversible refrigerator such as that depicted in Figure 9(f). To answer this question all you have to do is to reverse the directions of all the flows shown in 8(d). Once this is done you'll realize that this work would be capable of sending Q_c^* (as well as itself) back to the hot reservoir. At the end of this process, however, there still remains Q_i to be restored (remember that $Q_c = Q_c^* + Q_i$) and no work left to carry on this restoration. The restoration of this amount of heat demands that the work required be imported from somewhere else. Under this scheme of things the restoration of Q_c not only consumes the work W originally generated in 8(b), it also consumes an extra amount of work for the restoration of Q_i.

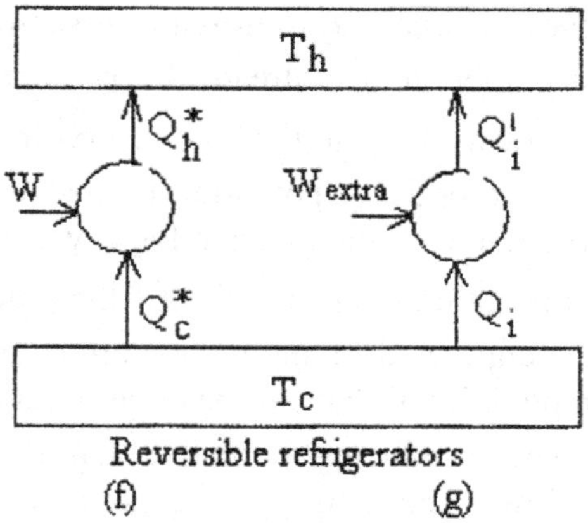

Figure 9. Reversible refrigerators used to carry on the restoration of Q_c^* and Q_i. Refrigerator 9(f) is the inverse of engine 8(d)

At the light of these results Thomson appears to be correct: any restoration of mechanical energy without further dissipation is impossible. The gradients available for the generation of mechanical energy are disappearing. The price for keeping these gradients intact so the same amount of mechanical energy can be obtained from them at any given time can only be done at the price of spending in their maintenance amounts of work $(W + W_{extra})$ larger than that they are capable of producing in the first place (W). Bad business this is indeed. If the original process 8(b) had been reversible, Q_i would have been zero. In this situation the work produced by 8(b) would have been identical to that required by a reversible refrigerator to produce the restoration of Q_c to the hot reservoir. It is only under reversible conditions that on Thomson's perspective we can break even, as it is only here that the maintenance work is identical to that produced by the cycle.

This scenario, even if reasonable at first sight, is however, flawed. In order to locate the flaw in Thomson's argument we have to remember that Q_c subsumes two radically different heat transfers: Q_c^* flowing reversibly, and Q_i flowing irreversibly. It has already been mentioned in this regard that the flow of Q_c^* to the cold reservoir carries with it no unused work producing potential, no lost work. If this is so then, in the name of elementary logic we would have to conclude that if nothing is lost with Q_c^*, then there is nothing from it to be recuperated. If any recuperation, any restoration is to be attempted, it will have to be done in regard to Q_i and the unused work-producing potential that it carries in its irreversible flow. The entropy change associated to this irreversible transformation is indicating us that in its transit to the cold reservoir, this heat could have delivered an amount of work equal to $Q_i(T_h - T_c)/T_h$. It however, hasn't. Since this is the only mechanical energy lost, this is the only mechanical energy that can be recuperated. Otherwise we will be confusing the orange with its peel.

With this conceptual correction in place it follows that in the case of a reversible engine there is no mechanical energy to be restored because none was lost in the first place. The whole of the work there produced remains as, say, 'surplus work'. The general problem of restoration, can then be stated in reference to Figures 8 and 9, as follows: At what efficiency should process 8(b) operate in order for the work by it outputted be capable of precisely producing, when fed to process 9(g), the restoration of Q_i? In other words, what is the efficiency at which 8(b) has to operate for $W = W_{extra}$? The solution of this problem starts by writing the following equation for reversible refrigerator 9(g)

$$W_{extra} = Q_i \frac{T_h - T_c}{T_c} \qquad (8.21)$$

As previously stated in regard to Figure 8, we have that

$$Q_i = Q_h - Q_h^* \qquad (8.22)$$

Substitution of eq. (8.22) in (8.21) leads to

$$W_{extra} = Q_h \frac{T_h - T_c}{T_c} - Q_h^* \frac{T_h - T_c}{T_c} = Q_h \frac{T_h - T_c}{T_c} - W \frac{T_h}{T_c} \qquad (8.23)$$

The substitution performed in the last term of eq. (8.23) follows directly from Figure 8(b). Insertion in eq. (8.23) of the condition guiding this argument $W = W_{extra}$, produces, after some algebra, the following equation:

$$W(\frac{T_h + T_c}{T_c}) = Q_h(\frac{T_h - T_c}{T_c}) \qquad (8.24)$$

Recognizing the quotient W / Q_h as the efficiency of the operation fulfilling the said condition, we can finally write:

$$\eta_{umbral} = \frac{T_h - T_c}{T_h + T_c} \qquad (8.25)$$

The already proven fact that $\eta_{rev} \rangle \eta_{umbral}$, allows us to realize that any process with an efficiency η such that $\eta \rangle \eta_{umbral}$ will produce amounts of work W larger than W_{extra}, and if so, that once this 'maintenance' of our energy transforming machinery has been taken care of, we will still have an amount of surplus work $W_{surplus} = W - W_{extra} \rangle 0$ for the labor of progress, of growth, of development. This is the correct perspective to Thomson's problem of restoration of mechanical energy.

The '*umbral*' name given to this efficiency obeys to the fact that it is identical to that shown in eq. (8.15). In the case here being considered this efficiency separates those operations producing enough work ($\eta \geq \eta_{umbral}$) to bring about restoration from those that doesn't ($\eta \langle \eta_{umbral}$).

If a non-reversible refrigerator, say a refrigerator of efficiency comparable to that of the engine 8(b) is used to produce restoration, then the engine will have to produce enough work to restore not only the work lost in its own operation, but that lost in the operation of the refrigerator as well. This efficiency, which can be called 'the ultra efficiency', can be determined through a thermodynamic analysis similar to that leading to equation (8.25). Even if it is possible to define and determine other efficiency thresholds, that will not be done here. In the never ending motion towards more efficient energy transforming machinery which characterizes human evolution, the crossing of a given efficiency threshold defines operations with an increased set of capabilities. The ability to restore the lost work coming from both, the engine and the refrigerator present at efficiencies larger than the *ultra* subsumes as well as surpass the abilities present at the lower stage of development characterized by efficiencies between the *umbral* and the *ultra.*

A discussion relative to the efficiency of couplings of heat engines relevant to the ideas previously introduced has been previously published by this author [Íñiguez, 2010]

8.4 THREE BODIES CHANGING, THREE ENTROPY CHANGES

In section 8.1 and in regard to equation (8.7) it was said that the negentropic effect of the heat to work transformation (or the entropic effect of the transformation of work into heat in a refrigerator) appears or manifests itself as a separate, distinct entropic effect from those of the reservoirs. At difference of Clausius formulation which considered the entropic effect of the heat to work transformation as subsumed by that of the hot reservoir, here we can see that it is actually independent of it.

Ask yourself now the following question: How many bodies are affected by one cycle in the operation of a reversible engine? If your answer is: two, you are wrong! The correct answer is: three. At the end of one cycle, alongside the changes experienced by the two heat reservoirs, we also find the mechanical reservoir in a condition different from the one it had at the beginning of the process. As a consequence of it safekeeping the work produced –assuming the mechanical reservoir is of the pulley/weight kind- we would find that the weight has increased its height to a new level proportional to the amount of work produced by the cycle. The conclusion is simple: three bodies changing means three entropy changes. The material constitution, the form, and the mode of operation of the mechanical reservoir are actually immaterial; inconsequential to this analysis. What matters is that its change is the *tangible* manifestation of the transformation of heat into work. At the light of these considerations we can write:

$$\Delta S_u[Cycle]_{rev} = \Delta S[hot\ reservoir] + \Delta S[cold\ resrvoir] + \Delta S[mech.\ reservoir]$$

(8.26)

Making use of the fact that is through the change in the mechanical reservoir that the heat to work transformation manifests itself –the reader can now take a fresher look at Barrow's quote in section 5.1- we will have that

$$\Delta S_u[Cycle]_{rev} = -\frac{Q_h}{T_h} + \frac{Q_c}{T_c} + \Delta S[Q(T_h) \rightarrow W]_{rev} \qquad (8.27)$$

And also that

$$\Delta S_u[Cycle]_{rev} = -\frac{Q_h}{T_h} + \frac{Q_c}{T_c} - \frac{Q}{T_h} \qquad (8.28)$$

This is the rationale behind equation (7.26).

8.5 THE NOTIONS REPLACING THE LAW OF INCREASING ENTROPY

It is indeed true that as Planck states "…there exists in nature a quantity which changes always in the same direction in all natural processes." [Planck, 1990, pp. 106] Contrary to common dogma, however, this quantity is not the entropy change for the universe of the process being considered, but the always positive change of its irreversible component. Therefore, if an entropy-centered statement of the second law is to be voiced it should identify the said always increasing quantity with the entropy change of the irreversible component and not with the universe entropy change. An equivalent statement to the one previously annotated can be voiced by saying that it is impossible to revert the effects of the irreversible component of any process, without changes being left somewhere in the universe, i.e. without compensation.

The irreversible component of every natural process subsumes entropy producing, dissipative, work-degrading processes. The fact that any of these dissipative processes can be reduced to the direct flow of an amount of heat from a hotter to a colder body, combined with the fact that the transfer back of this heat from the colder to the hotter body cannot be produced without compensation allows us to understand that the essence of irreversibility, as originally recognized by Clausius, boils down to the prohibition nature's imposes on the spontaneous, unassisted transfer of heat from colder to hotter bodies.

Now, If it is true that the significance of the second law of thermodynamics "…depends on the fact that it supplies a necessary and far- reaching criterion as to whether a definite process which occurs in nature is reversible or irreversible" [Planck, 1990, pp. 86-87] then, in recognition of the fact that irreversibility is a consequence of the inability of heat to flow unassisted from a colder to a hotter body, the essential statement of the second law should be no other than the following

Heat cannot, of itself, pass from a colder to a hotter body [Clausius, 1879, p.78]

8.6 THE END OF CERTAINTY IN SECOND LAW THERMODYNAMICS

The notion that the entropy of the universe tends to a maximum has transited over the years from a physical 'law' to nothing short of a religious dogma. The statement of Eddington seen from the perspective of Saint Augustine's quote of chapter 5 illustrates well the position of most thermodynamicists in this regard. This situation is explainable once we realize that the law of increasing entropy not only provides certainty and guidance in regard to the direction all natural processes follow; Perhaps more important, it also answers with authority, detail, and physical 'evidence' the question about the ultimate fate of the universe, a question for which no religion has an answer of comparable certitude. Looking at it from this perspective it is not that strange that this law had developed a religious smell, or that the 'sacred cows' of second law thermodynamics had become -for all practical purposes- the supreme priests of its temple, with the sacred duty of *a priori* rejecting any sacrilegious position doubting the universal validity of their creed.

The evidence given indicates, however, that the entropy is a gauge for the effects brought upon not only by the transformation of work into heat, but also for the transformation of heat into work, as Clausius originally recognized. In reference to heat engines the evidence also indicates that the reversible limit is not isentropic but negentropic. Through this evidence we've been able to witness the domain of validity of the certainties offered by the law of increasing entropy reduced, *strictu sensu,* to the irreversible limit ($\eta = 0$). At the reversible limit ($\eta = \eta_{rev}$) this law has actually been replaced by its opposite: *the law of decreasing entropy.* In any other situation i.e. for any partially reversible cyclical process with an efficiency of operation bound by those corresponding to these two extremes, the total entropy change can be either, positive, negative, or zero, depending on how this efficiency compares with the umbral efficiency.

The results here offered bring forward the realization that in place of the unjustified certainty offered by the law of increasing entropy in the form of the heat death of the universe, a position subsuming the impotence, the irrelevance of the journey of man on this Earth as well as his inevitable subjugation to forces beyond his control; we find that the fate of our world is determined, as equation (8.19) makes evident, by the competition of two opposite effects: Chaos vs. Order or Entropy vs. Negentropy, embodied these in the terrestrial, human-centered notions of Gained Work vs. Lost Work, on whose resultant the contribution of the human effort is determinant. The future of our Earth and of the human species with it depend no more on the action of quasi-metaphysical, eschatological forces; it actually resides on the working hands and creative minds of men.

CHAPTER 9: CLAUSIUS' ANALYSIS, THE LAW OF INCREASING ENTROPY, AND THE NEGENTROPIC FORMULATION AT THE LIGHT OF DIALECTICS

The notions furnished by Clausius, current thermodynamics, and the negentropic formulation will be, in what follows, analyzed at the light of each of the three laws of dialectics.

9.1 THE LAW OF THE TRANSFORMATION OF QUANTITY INTO QUALITY

The efficiency of a heat engine quantifies the fraction of every unit of heat that ends up being transformed into the superior form of energy we call work. As the knowledge of mankind progresses so does the efficiency of its energy transforming machines. At increased levels of efficiency the proportion of gained to lost work increases, as does the negentropic contribution of the former to these engines's universe entropy change. The continuous quantitative additions of 'efficiency' to the universe of our energy transforming machines made possible by the advance of technology leads, inevitably, to efficiencies larger than the *umbral* efficiency. Once this threshold is breached a qualitative change in their universe takes place in the form of an entropic to a negentropic transition and with it, the acquisition on part of these universes of capabilities not present within the entropic domain. These engines are now capable of at least restoring the loss of work associated to its operation. If this addition of 'efficiency' is continued the universe will be able to cross another threshold represented now by the *ultra efficiency*. The new capability associated to efficiencies larger than the *ultra* consists in the capacity of these engines of generating enough work to restore the lost work associated to its operation as well as that produced by the non-reversible refrigerator used in the restoration process. In the negentropic formulation, in a manner that accords with the first law of dialectics, quantitative additions of 'efficiency' bring forward qualitative transformations in the universe of these processes, expressed in a new and increased set of capabilities.

From the old formulations we learn that the entropy change for the universe of a heat engine transits from a positive value of $Q_h(T_h - T_c)/T_hT_c$ for an efficiency of zero, to a value of zero for the reversible efficiency. The fact that reversible processes are actually ideal processes, produce as a result that in the old formulations the entropy change of any actual process, whatever the value of its efficiency might be, is always positive. The possibility of a negentropic universe is actually alien to these constructions, as is the concept of *umbral* efficiency. If so, no entropic-negentropic transition is at all possible for the universe of these processes, nor access to the newer and superior capabilities associated to the negentropic domain. In the old formulations any possibility of progress, understood as the access to qualitatively superior states, is denied. The future here is not represented by progress, but by regress; a slow but inexorable fall into the final equilibrium condition represented by the heat death of the universe!

9.2 THE LAW OF THE INTERPENETRATION OF THE OPPOSITES

In the universe of heat-work inter-conversions i.e. in the universe of heat engines we find two opposites at play: On one extreme we find the zero efficient, irreversible, work loser ergo entropic operation represented in Figure 7(a); and in the other the maximally efficient, reversible, work gaining ergo negentropic operation shown in 7(c). As true opposites, their respective 'entropy' contributions are of opposite sign.

The varying degree of prevalence of one of these opposites over the other -measured either by the reversibility degree or the efficiency of the operation- determines not only the magnitude, but also the sign of the entropy change to such an operation associated. When the inter-penetration is such that the entropic contribution of the irreversible opposite is equilibrated by the negentropic contribution of the reversible opposite we find the operation taking place at constant entropy; here the efficiency of the operation corresponds to that which had been called the umbral efficiency. If the progress of mankind can be tied to the increased efficiency of our energy transforming machines, it can then also be tied to a steady motion in the negentropic direction. On this perspective the arrow of progress, of evolution, and consequently of time, point in the negentropic direction.

In current thermodynamics the negentropic contribution of heat-to-work transformations is negated; and in doing so also negated is the law of the interpenetration of the opposites. In current thermodynamics there is only one force in operation, only one source contributing to the universe entropy change: the irreversible component. The fact that in heat engines this contribution is none other than that associated to the direct, irreversible flow from the hot to the cold reservoir of the $(1-\phi)$ fraction of the incoming heat Q_h allows us to realize that current thermodynamics considers the 'engine of nature' as having one and only one capability: the dissipation of gradients. This is the only path available to our universe in the eyes of current second law thermodynamics, a path whose inescapable destination is the heat death of the universe. In *de facto* negating the existence of the reversible opposite, this formulation blinds itself from the fact that it is work, in all of its manifestations, the one that embodies the possibility of motion in the opposite direction i.e. the one that lifts a weight, that winds a clock, the one that brings progress and through it the access to increased efficiency levels, the one that in the proper circumstances created, maintains and propels the evolution of the astonishing diversity of all these interconnected self-organizing structures –man included- that has colonized our planet and that on reason of its intrinsic creative power constitutes the opposite to the dissipative, destructive forces of irreversibility. It is only through the interplay of these two opposites: chaos and order, entropy and negentropy, progress or regress, lost work and gained work, that a coherent image of the being and becoming of our world can be constructed. Whether our world, our Earth and the human societies by it sheltered move in one direction or the other depends on the degree of interpenetration of these opposites.

9.3 THE LAW OF THE NEGATION OF THE NEGATION

Any actual work-producing cyclical process ($0 \langle \eta \langle \eta_{rev}$) represents the negation of the workless, inefficient, and entropic irreversible limit. As true negations these cyclical processes carry with them, alongside the work producing capability defining their superior stage of development, the imprints of the previous stage in the form of an irreversible component. Each higher efficiency cyclical process is, in this perspective, the negation of the previous less efficient stage: the negation of the negation. The continuous transit along this efficiency increasing path is what defines the essential outcome of the law of the negation of the negation: progress, development. Materializing through increased levels of penetration of the reversible over the irreversible component, development takes form as the incessant higher-order re-expression of those quantities, relations, abilities, and laws characterizing the lower or previous level; with each new level representing a qualitatively different condition, a qualitative advancement over its predecessor. Thus understood development embodies the opposite notions of advance and regression, of progress and retrocession, of folding and unfolding, of entropy and negentropy. Starting with the original departure from the irreversible limit, development leads the universe of our energy transforming machines into a series of nested thresholds, the first of which, represented by the umbral efficiency, marks the transition from an entropic to a negentropic universe. As these thresholds are crossed, newer and increased capabilities are accessed, each of these representing the repetition at a higher stage of development of the capabilities of the lower stage. Just like in heat engines the *umbral* and *ultra* stages can be recognized, in human development these thresholds have been represented by the so-called 'ages': the stone age, the bronze age, etc., each one representing a set of knowledge, techniques, abilities, organization, and most of all an efficiency in the utilization of the human effort recognizably different, markedly superior from the previous one. Transit in the opposite direction, it should be clear, is regress, as it implies lesser and lesser efficiencies and with it the losing of previously acquired capabilities.

In having access only to changes in condition characterized by positive entropy changes, for the old formulations the possibility of accessing qualitatively different states is non-existent. The three dimensional, ever-growing, ever-expanding helix of progress defined by dialectics –the same we've found subsumed by the negentropic formulation- becomes a spiral in both Clausius and current thermodynamics. Confined in a sterile two dimensional surface, every change in their universes taking place is directed towards a central point –equilibrium- in which all change is to stop. At the eyes of the theories of self-organization that find sustenance in the old formulations, the constant development taking place within and around us are nothing but the blind actions of a self-destructing and stupid universe eager to dissolve as soon as possible in the *Gotterdammerung* of its heat death; as according to these theories every self-organized structure takes form and develops at the price of increasing the rate of entropy production [Prigogine, 1980].

9.4 CONCLUDING COMMENTS

In Clausius construction the law of increasing entropy has been shown to be nothing more than a good example of bad science. In current second law thermodynamics the evidence suggest it to be the product of skewed, manipulated science. We have had here the opportunity to give this law its real dimension. From the position as supreme law of nature, it has been reduced to a law valid only at the lifeless, desolate, irreversible limit on which no work at all comes out of the available gradients. At this end of the efficiency spectrum all processes reduce to the irreversible transfer of heat from high to low temperature, this is, to the destruction of gradients, and because of this here the 'heat death' is the only future possible. Had this been the only option for our universe, no life could have emerged, nor developed. Life, in its very essence, is one of the manifestations of the work producing capacity of this engine we call universe. The same goes for development, progress, and advancement. Aren't they the manifestation of the increased work producing capabilities – the energy density output- of societies?

The arrow of time is not, can not, be entropy production. Entropy production points towards the inefficient, work degrading, irreversible extreme. If time points in the direction of evolution, of development, then it points in the direction of increased efficiencies not only in the production, but also in the utilization of ordered energy, of work.

We have proved here that the total entropy change of reversible processes can be negative, as for $[Q(T) \rightarrow W]_{rev}$; positive, as for $[W \rightarrow Q(T)]_{rev}$; and zero, as for $[Q(T) \rightarrow Q(T)]_{rev}$. We have also proved that irreversible cyclical processes working with efficiencies smaller than the *umbral* are entropic; those working with efficiencies larger than the umbral but smaller than the reversible efficiency are negentropic; while those working at precisely the umbral efficiency are non-entropic. This new knowledge indicates that total entropy is certainly not what we were thought to believe. It is not an irreversibility indicator for as we have seen, it does not remain constant for all reversible processes; it is not an indicator of the flow of time because it doesn't change in the same direction for all irreversible processes. The sign and magnitude of the total entropy change for a given thermodynamic process is nothing more and nothing less than an indicator of the prevalence of one of these opposites over the other. An entropic universe is the product of the dominance of work degrading over work producing. The opposite is true for negentropic processes. A perfect balance between work degrading and work producing means non-entropic universes; sort of what other disciplines call zero-growth. This condition means stagnation. Nor progress, nor regress. This balance point constitutes the ideal condition to certain kinds of mind.

Getting at the results herein presented demanded the recognition of certain notions that had been overlooked by the old formulations. The discussions offered in Figure 2' in regard to the effect that the degradation of the work produced through $[Q(T_h) \rightarrow W]_{rev}$ in a reversible cyclical process has on the condition and entropy change of its accompanying transformation $[Q_c(T_h) \rightarrow Q_c(T_c)]_{rev}$, as well as that introducing the '*W*' Transformation are essential to the new formulation. Their understanding demands, however, a prejudice-free attitude, a not so common trait according to my own experience, or that of Tolstoy's for that matter:

I know that most men, even those at ease with problems of the greatest complexity, can seldom accept even the simplest and most obvious truth if it be such as would oblige them to admit the falsity of conclusions which they have delighted in explaining to colleagues, which they have

proudly taught to others, and which they have woven, thread by thread, into the fabric of their lives.
http://en.wikiquote.org/wiki/*Leo_Tolstoy*

Even if I am almost positive that my message will not permeate the conceptual barriers and limitations of the thermodynamicists of my generation, I do have hope that the newer generations might adopt a different attitude:

A new scientific truth does not triumph by convincing its opponents and making them see the light, but rather because its opponents eventually die, and a new generation grow up that is familiar with it.
http://en.wikiquote.org/wiki/Max_Planck

REFERENCES AND NOTES

Aristotle, 350 BCE, *Metaphysics*, book IV, Parts 3 and 4; as translated by W. D. Ross, and posted on the Internet Classic Archive at MIT: http://classics.mit.edu/Aristotle/metaphysics.4.iv.html

Barrow, G. M., 1973, *Physical Chemistry*, 3rd. Ed, McGraw-Hill, New York

Caillois, R., 1973, *Coherences Aventureuses*, Editions Gallimard, Paris
The essential message of this quote from Roger Callois is that the positions of Clausius and Darwin are incompatible.

Carnot, S. 1960, *Reflections on the Motive Power of Fire and on Machines Fitted to Develop that Power*, Dover, New York

Césarman, E., 1997, *Termodinámica de la Vida* – publicación conjunta de las obras '*Hombre y Entropía*' y '*Orden y Caos*'- Ediciones Gernika S.A., México, D. F.

Clausius, R., 1879, *The Mechanical Theory of Heat*, MacMillan, London
This book is available at Google Books. (Accessed May 04, 2010)

Clausius, R., 1862, On the Application of the Theorem of Equivalence of Transformations to the Internal Work of a mass of matter, *Phil. Mag*. S. 4. Vol. 24, No. 159

DeChardin, T., 1967, La Activación de la Energía, Taurus Ediciones, Madrid
"Para abarcar completamente la economía evolutiva del Universo (incluyendo a la Vida), debe añadirse un tercer Principio, el de la Reflexión de la Energía, a los ya admitidos, de la conservación y de la degradación de la Energía"

Eddington, A. S., 1915, *The Nature of the Physical World* "The law that entropy always increases hold, I think, the supreme position among the laws of Nature…" http://en.wikiquote.org/wiki/Thermodynamics (Accessed May 03, 2010)

Engels, F., 1964, *Dialectics of Nature*, Third Revised Edition, Progress Publishers, Moscow

Engels, F., 1987, *Anti-Dühring* and *Dialectics of Nature*, Published jointly as Volume 25 of *Karl Marx and Frederic Engels Collected Works, Vol. 25*, Lawrence & Wishart Ltd., London, International Publishers Co. Inc., New York, and Progress Publishers and the Institute for Marxism-Leninism, Moscow.

Íñiguez, J. C., 1999, A Revision of Clausius Work on the Second Law, *Entropy*, Vol. 1, pp. 111-147

Íñiguez, J. C., 2008, *Negative Entropy: A brief incursion into the uncharted universe of decreasing entropy*, ISBN 0-7414-4898-X, Infinity Publishing Co., Conshohocken, Pa.

At the time of writing this book, as well as the work of the previous reference [Íñiguez, 1999] the author had not yet recognized the effect the degradation of W has on both, the nature and the entropy change of the transformation of Q_c shown in Figure 2' above. Even if most of the material presented in these two works retain its validity despite the situation mentioned, it is only prudent to say that all of my previous discussions on the subject of the second law have been superseded by the one contained in the present work. The argument unveiling the flaw existing within the formalism of current second law thermodynamics presented in chapter 6 is the matter of [Íñiguez, 2011]

Íñiguez, J. C., 2010, An Unusually Efficient Coupling of Carnot Engines, *Apeiron*, Vol. 17, No. 2, pp. 150-172

Íñiguez, J. C., 2011, A Thermodynamic Impasse: A Constant Entropy Irreversible Process, *Physical Chemistry: An Indian Journal,* Vol. 6, Issue 3, pp. 125-130

Klotz, I. M., Rosenberg, R. M., 1986, *Chemical Thermodynamics*, Fourth Edition, Benjamin/Cummings, Menlo Park, California

Maron, S. H., Prutton, C. F., 1965, *Principles of Physical Chemistry*, 4th Edition, Mac Millan, International Student Edition, New York

McGlashan, M. L., 1966, *J. Chem. Educ.*, Vol. 43, No. 5, p.232

MIA: Encyclopedia of Marxism; www.marxists.org/glossary/terms/l/a.htm#laws-dialectics (Accessed April 2012)

Pitzer, K. S., Brewer, L., 1961, *Thermodynamics,* 2nd Edition (Revision of Lewis and Randall) McGraw-Hill International Student Edition, New York

Prigogine, I., 1980, p. 89, *From Being to Becoming*, Freeman, New York

Of course, all of the thermodynamic, entropy-centered analyses that up to date have been produced on self-organizing systems ignore the work gained made evident in the structural and functional order of these systems. Only the lost work is taken into account. It is on reason of this that no validity can be ascribed to the conclusions on self-organized systems advanced by these old-formulations-based studies.

Schmidt, E., 1966, *Thermodynamics*, Dover, New York

Thomson, W., 1852, On a universal tendency in nature to the dissipation of mechanical energy, *Proc. Roy. Soc. Edin. III*

Made in the USA
Monee, IL
05 November 2026

67385487R00077